QUANTUM PHYSICS

FOR BEGINNERS

The Ultimate Guide to Discover the Theory, Secrets, and Wonders of Science that Changes Your Life.

Easily Learn the Theories of Energy of Black Holes and Relativity

Rodney McPatterson

1

Table of Contents

Introduction

Simply described, it is physics that describes how it works—the strongest explanation we have of the nature of the objects that make up matter and the forces by which they communicate.

Quantum physics is the foundation behind how atoms work and why chemistry and biology work as they do. You, me, and the gatepost—at least at some point, we're all dancing to the quantum theme if you want to understand how electrons travel through a computer chip, how photons of light transform into electrical current in a solar panel or enhance themselves in a laser, or just how the sunburns, you'll need to use quantum mechanics.

The complexity—and, for the physicists, the pleasure—starts here. First of all, there is no particular quantum theory. There is quantum mechanics, the fundamental mathematical structure that underpins all that was first formulated in the 1920s by Niels Bohr, Werner Heisenberg, Erwin Schrödinger, and others. This characterizes basic stuff like how the position or momentum of a single particle or a community of several particles varies over time.

Yet to clarify how things function in the physical universe, quantum mechanics must be coupled with certain aspects of physics—primarily, Albert Einstein's special relativity principle, which describes what occurs as objects pass quite quickly—to establish what is regarded as quantum field theories.

Three separate quantum field theories deal with three of the four basic forces that matter deals with; electromagnetism, which describes how atoms bind together; strong nuclear power, which describes the structure of the nucleus at the center of the atom; and poor nuclear power, which explains why certain atoms are prone to radioactive decay.

Billions and billions of experiences in these noisy conditions allow for the creation of "powerful field concepts" that gloss over some of the gory details. The challenge in building such theories is why many essential questions in solid-state physics remain unanswered—for example, why at low temperatures certain materials are superconductors that require current without electrical resistance and why we can't get this trick to operate at room temperature.

Yet, there is a massive fundamental mystery underneath all these functional issues. At the simplest stage, quantum mechanics proposes very peculiar things regarding how matter functions, which are entirely at variance with how things appear to work in the modern world. Quantum particles may behave like electrons distributed in a single place, or they can behave as waves, dispersed across space or in many locations at once. That they appear to rely on how we want to quantify them, and until we calculate them, they seem to have no definitive properties at all—bringing us to a profound conundrum regarding the existence of simple truth.

This fuzziness contributes to obvious paradoxes, such as Schrödinger's tale, in which a cat is left dead and alive at the same time due to an unknown quantum mechanism. Yet that's not half of it. Quantum particles often appear to be able to impact each other instantaneously, particularly though they are far apart from each other. A genuinely bamboo-like phenomenon is defined as entanglement or, in a term invented by Einstein (a true opponent of quantum theory), "spooky behavior at a limit." These quantum forces are unknown to humanity, and they form the base of new technologies such as ultra-secure quantum cryptography and ultra-powerful quantum computing.

Yet, no one understands what it all entails. Some people believe we only have to agree that quantum physics describes the natural universe to the point that we consider it difficult to fit our understanding of the broader, "classical" environment. Others believe there may be a stronger, more logical explanation out there that we are yet to find.

There are many elephants in the house in all of this. For example, there is a fourth basic force in existence that has so far been unable to describe the quantum theory. Gravity is the domain of Einstein's general theory of relativity, a distinctly non-quantum principle that does not even include particles. Massive attempts over decades to put gravity under the quantum umbrella and thereby clarify all basic phenomena inside one "theory of all" have come to nothing.

In the meanwhile, cosmological calculations suggest that more than 95% of the universe consists of invisible matter and dark energy, things about which we have little understanding inside the standard model, and conundrums such as the nature of the role of quantum mechanics in the chaotic workings of existence remain unknown. The universe is at a quantum stage—but whether quantum mechanics is the final word on the universe is an open question.

What Quantum Physics Is

For the vast majority of people, the term "quantum physics" is closer to "rocket science" than it is to "the wonders of the universe." And that's a real pity. Most of you might think of tedious formulae and explanations when thinking of physics—but the truth is that both "traditional" physics and quantum physics are pretty much the sciences that hold the secrets to the universe—the whys and the hows of the entire cosmos functioning.

No matter what you do are for a living, quantum physics will bring a whole new perspective into your life on so many things that it is impossible to ignore. How could you, when you know that quantum physics is at the foundation of what you are, in the background of your fate spinning your life, and at the core of your very way of "functioning" as an intelligent being of the universe?

Almost borderline between science and spirituality, quantum physics might finally be able to explain the unexplainable and help us transgress the borders of thinking that have been limiting us thus far and bring us closer to the essence of the world.

Let's dive in and uncover the basics of quantum physics!

What Is Quantum Physics

To understand what quantum physics is, you must first go to the "mothership," i.e., physics. For many people, physics is that boring subject in school you have to be a real "nerd" to like. The one that is even worse than mathematics and even more difficult to understand than chemistry.

For many other people, physics revolves around mechanics, or, in layman terms, "how cars work." While it is true that physics deals, among many other things, with how cars work, that it also deals with a bit more than how cars work. It is also worth noting that mechanics is only one branch of physics but lies at the very foundation of car-making, alongside electronics—another physics branch.

The etymology of the word "physics" is pretty fascinating. It comes from the Greek word physique, which is used to mean "knowledge of nature." As such, the definition of physics is tightly connected to nature and getting to know it. Many define physics as a natural science that studies matter, how it behaves through space and time, and how it connects to energies and forces.

Each of these deals with a different aspect of matter and other types of value (such as in nuclear physics, for example, which studies how atomic weight behaves in different contexts).

In addition to this categorization, you may also find people talking about classical physics and modern physics, which is a way to look at this natural science from the perspective of its evolution in time. So, where does quantum physics lie in this entire paradigm, you may ask?

Well, you see, quantum physics is a bit of an odd "animal" because it is sometimes used as a synonym for modern physics. So, it only comes as both a continuation of traditional physics and as an antagonist as well. Although most modern physics revolves around quantum theory, it is worth noting that, at large, it is still considered to be only a direction of modern physics. To better understand the relationship between modern physics and quantum physics, consider the fact that two significant theories and theoreticians marked the beginning of modern physics.

Planck's Constant is the most famous one, and it entailed that the energy and frequency of light are proportional, which led Einstein to postulate that light exists in small quantities of energy called "photons." Albert Einstein, whose major work was related to the theory of relativity and the photoelectric effect. The first postulates, in short, that gigantic objects can be the origin of a distortion in space and time, which is sensed as gravity. The latter says that light does not exist in waves but quanta (small pockets of energy), as mentioned above. If "big" physics deals with all the things, you can more or less see (or at least perceive, especially if you run experiments), quantum physics deals with the tiniest parts of matter. This is its actual definition; The science that deals with the atomic and subatomic levels of significance. That might not sound like much. However, quantum physics has gone so in-depth (and continues to do so) that it might be the one to explain everything in the universe finally. Everything we've never known. All the questions we always sought an answer to—the very essence of life. Can you think of quantum physics as a boring topic when you look at it from this perspective? When do you know that it is the science that will finally help us understand so much our place in the universe, more about ourselves, and where all of this is leading?

Not only does it appear to hold the key to all the things we never managed to achieve (like teleportation, or understanding fate and destiny, for example), but it is also a contradiction to a lot of very well-established theories, including that of general relativity brought forward by Albert Einstein himself. There is a war for knowledge out there, and quantum physics just "happens" to lie at the very core of it. We bet we made you curious!

Chapter 1.   The Difference Between Classical and Quantum Physics

Physicists are trying to understand the world—but to date, there has surfaced no proven, palpable theory to bring the two worlds together and finally help us understand where we come from, where we are, and where we are going (because, at the end of the day, these are the fundamental questions both classical and quantum physics propose).

In classical physics (as drawn out by Einstein's general relativity principle), this reality is made out of 4 dimensions (also called the space-time continuum). In this paradigm, gravitational fields are continuous entities.

In quantum mechanics, however, fields are not continuous but discontinuous. They are not defined by the four dimensions but by "quanta." As such, concepts like the "gravitational field" are missing from the world of quantum physics, which is also the biggest bridge classical physicists and quantum researchers have to build between their points of view.

This is not just a matter of fancy definitions. The world of quantum mechanics and the world of classical physics are incompatible because they describe reality in completely different ways, in different terms, and indifferent perspectives that do not meet at any point.

In classical physics, things happen for a reason. They happen according to the old cause-and-effect dictum. Nothing happens randomly, but because there is something else before it that has caused it.

In quantum physics, scientists do not see reality in terms of cause and effect but in terms of particles jumping from one state to another based on probability rather than definite outcomes.

Why is reconciliation important, then, especially given that these two disciplines seem so different and at such a deep level?

Because reconciliation would create a whole theory that explains the universe on a small and large scale, where classical physics fails to give explanations of the microcosms, quantum physics would succeed. And where quantum physics fails to make sense when it is blown up to macro-objects (remember the cat that was both dead and alive?), classical physics would be able to breathe in some logic.

Fundamentals About Classical Physics

Classical physics is invariant for time reversal, and we have seen that this gives us serious problems when we try to find an explanation for the thermodynamic arrow of time. It is, therefore, important to investigate now on dynamic reversibility in quantum mechanics.

Before we have a look at particle physics, quantum theory, and the cosmos, we need a brief introduction to the concepts of classical physics:

- Energy
- Weight and mass
- Matter—solids, and liquids
- Measures and units

Energy

Energy must be transferred to an object to perform work on or heat it.

Newton's law of the conservation of energy states that it may be transformed from one form to another. It cannot be created or destroyed. The SI unit for energy is the joule.

Forms of Energy

Types of energy include:

- Kinetic energy (movement)
- Chemical energy (e.g., coal, natural gas, etc.)
- Thermal energy
- Magnetic energy
- Light energy
- Electric energy
- Gravitational potential energy
- Nuclear energy

Weight and Mass

Weight is therefore dependent on the gravitational force where the object is situated. According to the International System of Units (SI Units), weight is measured in Newtons with the symbol N.

Matter

There are several forms of matter that we know of, but only three are of relevance to this book; solids, liquids, and gases.

Solids Liquids and Gases

- **Solids.** Solid objects have a defined shape because their atoms are packed tightly together (i.e., they have a high density). The atoms cannot move around and cannot be compressed into a smaller volume.

- **Liquids.** In liquids, the atoms are not so tightly packed, so they can flow around each other. Most liquids can be compressed into a smaller volume in a container, i.e., their atoms are forced closer together (they become denser).

- **Gases.** The atoms in gases are in constant movement and have a relatively large space between them. Gases can be compressed into a smaller volume when confined in a container and expand when released.

Measures and Units

- **Density.** The density of matter is a measure of how closely the atoms are packed together. Density is measured in kilograms per cubic meter and can be calculated by dividing an object's mass by its volume.

- **Volume.** The amount of space that matter occupies. There are many ways of measuring volume, depending on whether you measure solids, liquids, or gases. The formulae for measuring different shapes of solids or containers of liquids or gases are different.

As I promised in the introduction not to burden you with mathematics, other than E=mc2, I will leave you to find these formulae for yourself in further reading if you are interested. There are, however, some easy ways to measure small amounts of gases.

The units in which volume is measured can be confusing. The customary system in the US differs from the imperial system in the UK, and both of those differ from the metric system. In 1960, an international system of units was introduced. Le Systeme International Duties, now known simply as SI Units, are used by the scientific community to avoid confusion.

Fundamentals of Quantum Physics

The Quantization of Light

This was a forward step taken by Albert Einstein in the year 1905. With him, he suggested that quantization did not just involve mathematical tricks. On it, he added that it also involved the beam of light energy that is in the individual packets—currently referred to as photons. Thus, the energy of a single photon will be given by the product of the frequency of the energy and Planck's constant.

In the 19th century, the light was considered to be flowing in a wave, and this was the result of behaviors of light like polarization, diffraction as well as refraction. Actually, according to James Clerk Maxwell, magnetism, light, and electricity are all manifested by the same phenomenon, which is the electromagnetic field. He explains light as waves that constitute a combination of magnetic fields as well as oscillating electric. Einstein's "Photon model" came into place when it was able to explain the photoelectric effect successfully. This effect has been explained in the next step.

The Photoelectric Effect

According to Einstein's explanation, he argued that a beam of light had got the photons, which are particles that stream and also a frequency "f." The energy present in that photon will be equal to "hf." This implies that there is no effect on the energy that relates to the beam's intensity. He further explained that to remove an electron from a given metal, "work function" which is a certain energy amount is required—it is denoted by "φ." With his further explanation, when the work function is higher than the photon's energy, there will be no sufficient energy required to remove the electron from the given metal.

His description also argued that light is composed of particles that gave an extension of Planck's notion. This is the notion of energy that is quantized—whereby more or less amount of energy can be delivered by a given photon depending on its frequency. There was a compromise on the particle state of light since it was explained that light also had waves. This resulted in the consequences of the quantization of light.

Matter Quantization

According to Bohr, the electrons jumped from one orbit to another. In this case, it gave off the light emitted in the form of a photon. The energies that are emitted by the photon were highly dependent on the differences in energy that was present between the orbits. In the first place, there were very many critiques of Bohr's model. Many argued that this model was wrong, although, in the end, it was evident that the model was good to suit quantum physics. With his explanation, Bohr argued that matter has also got some wave-like properties. According to him, an electron beam can also exhibit "diffraction" This is a similar case, just like the beam of light or a wave of water. Thus, small molecules and atoms have got the same phenomenon.

Chapter 2.   The Achievements of Quantum Mechanics

In any case, a pleasant aspect concerning working in material science is that it's the most generally tried physics in humanity's set of experiences. There are delightful, transparent investigations showing the entirety of the peculiar properties of quantum material science. There are still a few contentions in the quantum establishment networks about how best to decipher what's "truly" proceeding to prompt the outcomes. However, the experimental proof is wholly unambiguous and without discussion.

Here are three of the many, numerous analyses demonstrating clear evidence that quantum material science is genuine, even the forecasts that appear to be genuinely peculiar.

Single-Molecule Interference

One of the soonest and most interesting forecasts of quantum material science is the possibility of molecule wave duality that everything known to man has both molecule and wave nature. Einstein was the first to truly push this, clarifying the photoelectric impact as far as what we presently call photons; Robert Millikan's trial of Einstein's physics demonstrated that it works splendidly made sure about Nobels for both Einstein and Millikan. The idea of material articles having wave nature came in 1923 from Louis de Broglie. The wave idea of the electron was before long shown by the Davisson-Germer try diffracting electrons off nickel (a cheerful mishap), and George Paget Thomson's diffraction of electrons off slim movies (fun certainty: G.P. Thomson's dad, J.J. Thomson, won a Nobel for demonstrating the electron is a molecule, at that point GP mutual one for showing it's a wave...).

Nowadays, the wave idea of the issue has been exhibited on many occasions, by and large, by indicating obstruction between issue waves going through at least two cuts made in a boundary. On the most distant side of the border, the waves meddle with one another to create an example of splendid and dim spots. Some time ago, Richard Feynman broadly said that obstruction of particles catches the real puzzle of quantum material science; at that point, this was still generally a psychological study; however, in the mediating fifty years, the specific test he talked about has been done several times, with multiple particles.

Thus, indeed, this single investigation contains all that there is to show indisputably that the wave idea of the issue is a genuine wonder.

Quantum Non-Locality

One of the most insightfully upsetting plans to originate from quantum material science is the way that physics is non-nearby. The results estimations made in inaccessible areas could correspond with one another in manners that would be outlandish if the assessments were autonomous. Data about the outcomes could go from one to the next at speeds slower than light speed. Einstein's last truly incredible commitment to material science was a 1935 paper with Boris Podolsky and Nathan Rosen spreading out the results of such a "trap" generally unmistakably. The essential thought is clarified in other materials.

This was viewed as an odd philosophical reference for quite a long time—Abraham Pais, remarkable logical history of Einstein gives it just several expendable passages. In 1965, however, the Irish physicist John Bell brought up that such a physics certainly preferred by Einstein, Podolsky, and Rosen set cut-off points on the sorts of connections you could hope to see between far off estimations, and those cut-off points were unique with what you could anticipate from an ensnared quantum framework. This roused various exploratory tests (empowered by Clauser, Horne,

Shimony, and Holt stretching out Bell's work to more functional conditions), all of which have affirmed that quantum mechanics does, truth be told, disregard the cut-off points forced by a "nearby shrouded variable" physics of the sort supported by Einstein and friends.

Regarding factual vulnerability, the absolute best tests have been made as of late, using connected photon sources that produce sets of photons and super high-productivity single-photon finders. The tests that persuaded most physicists, this was a genuine article and worth critical exertion, it was done in the mid-1980s by Alain Aspect and France's partners. They used calcium iotas eager to a specific high-energy state, from which the particles rot by discharging two photons (one red, one blue). At the point when those two photons are produced in inverse ways, their polarizations are caught in precisely the correct manner to test Bell's physics.

The preliminary test by Aspect used a solitary finder with a polarizer before it on each nuclear pillar side. It estimated how frequently they recognized photons at the two indicators for different mixes of the polarizer settings. The subsequent estimation discovered relationships that abused the Bell/CHSH limit for nearby concealed variable hypotheses. Yet, there was an escape clause because of the limited proficiency of the identifiers. The way that the identifiers now and then "miss" a photon implied that you could clarify their outcome using a neighborhood physics that coincidentally missed photons in an incredibly fortunate (or unfortunate, contingent upon which result you like) design. In this way, they rehashed the test with four identifiers, two on each side, and possibly checked information when they got one photon on each side. Once more, the relationships they estimated surpassed as far as possible by crazy multiple times the real vulnerability in their estimation.

However, there's a proviso here because they set the polarizer's points for their indicators and left them set up.

This opens such a paranoid fear proviso—a sign could go between the indicators and the source at the speed of light, mentioning to the start what the locator settings were, so, all in all, the identifier could decide to convey photons whose polarizations imitate the quantum expectation, however, are characterized by a neighborhood shrouded variable physics. So, Aspect did a third test, distributed in 1982, including a couple of quick switches that viably changed the polarizer settings while the photons were in flight. Once more, the outcomes disregard the cut-off points for nearby shrouded variable hypotheses, and such that shows the polarization in flight should be uncertain.

The Aspect tests are not authoritative in a manner that fulfills bad-to-the-bone thinkers—it's as of late that really "escape clause free" trials have begun to be practical (counting some that depend on cosmology to guarantee the autonomy of the indicators). Nonetheless, they are sufficient to persuade most physicists that non-area is a genuine marvel that should be paid attention to. What's more, that set off a decent arrangement of test and hypothetical exertion investigating how these functions, how to accommodate trap with relativity, and how to misuse that. This has prompted the formation of the blasting subfield of quantum data science, which produces innovative stunts like quantum teleportation and pragmatic advances like quantum cryptography.

The Aspect tests show (decently) indisputably that quantum material science is non-neighborhood and that the universe is a lot more abnormal than it shows up, or than Einstein would've enjoyed it to be.

Exactness Measurement

The movement of a solitary electron collaborating with an electromagnetic field appears to be something that should be truly easy to clarify, and the quantum physics of 1930 nearly hit the nail on the head. However, finding the correct solution requires disregarding a few impacts that everyone in theoretical material science knew should be genuine, which just worked until new tests empowered by WWII innovation made them challenging to overlook. The well-known Shelter Island Conference in 1947 tossed these issues into distinct help and started a significant exertion to comprehend them. After a year, at the Pocono Conference, Julian Schwinger and Richard Feynman had tackled the issue and created QED working models; Sin-ItiroTomonaga in Japan had his variant at a similar time, and Freeman Dyson indicated that each of the three renditions was numerically identical.

Feynman's rendition of physics is the easiest to use. This way, the most renowned—most present-day depictions of Schwinger's belief wind up depending on Feynman's language to discuss what's happening. It presents the possibility of "virtual particles" flying into reality out of void space, interfacing with the real particles whose properties we measure for a brief timeframe, at that point disappearing once more. This gives a helpful theoretical clarification to what's happening and bridles the intensity of account to make indeed extract science fathomable.

The Kilogram Goes Quantum

Another 2019 quantum expression point originated from the universe of loads and measures. The standard kilogram, the physical item that characterized the unit of mass for all estimations, had for some time been a 130-year-old platinum-iridium chamber weighing 2.2 lbs. what's more, sitting in a room in France. That changed for the current year.

The old kilo was great, scarcely changing mass throughout the long term. In any case, the new kilo is great. Based on the principal connection between mass and energy, just as a character in the conduct of life at quantum scales, physicists had the option to show up at a meaning of the kilogram that won't change at all between this year and the finish of the universe.

Reality Broke a Little

A group of physicists planned a quantum explore that indicated that realities change contingent upon your perspective on the circumstance. Physicists played out such a "coin throw" using photons in a little quantum physics, finding that the outcomes were distinctive at various identifiers, contingent upon their viewpoints.

"We show that, in the miniature universe of molecules and particles that is administered by the bizarre principles of quantum mechanics, two unique eyewitnesses are qualified for their realities," the experimentalists wrote in an article for Live Science. "As such, as indicated by our best physics of the structure squares of nature itself, realities can be abstract.".

Snare Got Its Excitement Shot

Unexpectedly, physicists made a photo of the marvel Albert Einstein depicted as "creepy activity a way off," in which two particles remain genuinely connected regardless of being isolated across separations. This element of the quantum world had been tentatively checked for some time, yet this was the first occasion when anybody got the opportunity to see it.

Something important went in numerous ways.

Somehow or another, the theoretical inverse of entrapment, quantum superposition, is empowered by a solitary item to be in (at least two) positions in one place, an outcome of an issue existing as the two particles and waves. Commonly, this is accomplished with minuscule particles like electrons.

In any case, in a 2019 investigation, physicists figured out how to pull off superposition at the most significant scale ever, using lumbering, 2,000-particle atoms from the universe of clinical science known as "oligo-tetraphenyl porphyrins improved with fluoroalkylsulfanyl chains.".

Warmth Crossed the Vacuum

Under typical conditions, warmth can cross a vacuum in just a single way, like radiation. (That is what you feel when the sun's beams cross space to beat all over on a late spring day.) Otherwise, in standard physical models, heat moves in two habits. First, invigorated particles can thump into different particles and shift their energy. (Fold your hands over a warm cup of tea to feel this impact.) Second, a warm liquid can uproot a colder drink. (That is the thing that happens when you turn the warmer on in your vehicle, flooding the inside with warm air.) So, without radiation, heat can't cross a vacuum.

Yet, quantum material science disrupts the guidelines. In a 2019 test, physicists exploited how vacuums aren't genuinely vacant at the quantum scale. Instead, they're loaded with little, irregular changes that fly into and out of presence. The analysts discovered that warmth could cross a vacuum by bouncing at a little enough scale, starting with one vacillation then onto the next over the unfilled space.

Circumstances and logical results may have moved in reverse.

This next finding is a long way from a tentatively confirmed disclosure, and it's even well external to the domain of conventional quantum material science. In any case, specialists working with quantum gravity—a hypothetical development intended to bring together the universes of quantum mechanics and Einstein's general relativity—indicated that an occasion might cause an impact that happened before in specific situations time.

Certain too weighty items can impact the progression of time in their prompt region because of general relativity. We realize this is valid. What's more, quantum superposition directs that an item can be in numerous spots without a moment's delay. Put a too bulky item (like a significant planet) in a quantum superposition condition, the analysts composed, and you can plan crackpot situations where circumstances and logical results occur out of order.

Chapter 3.   Quantum Theory

Max Planck, the Father of Quantum Theory

All objects emit electromagnetic radiation, which is called heat radiation. But we only see them when the objects are very hot. Because then they also emit visible light. Like glowing iron or our sun. Of course, physicists were looking for a formula that would correctly describe the emission of electromagnetic radiation. But it just didn't work out. Then, in 1900, the German physicist Max Planck (1858—1947) took a courageous step.

The emission of electromagnetic radiation means the emission of energy. According to the Maxwell equations, this energy release should take place continuously. "Continuously" means that any value is possible for the energy output. Max Planck now assumed that the energy output could only take place in multiples of energy packets, i.e., in steps. That led him to the correct formula. To the energy packets, Planck said, "quanta." Therefore, the year 1900 is regarded as the year of birth of the quantum theory.

Important. Only the emission (and the absorption) of the electromagnetic radiation should occur in the form of quanta. Planck didn't assume that it was composed of quanta because that would mean that it would have a particle character. However, like all other physicists of his time, he was persuaded that electromagnetic radiation was comprised of waves. Young's double-slit experiment has revealed it, and the Maxwell equations have established it.

In 1905 an interloper named Albert Einstein was much bolder. He took a glance at the photoelectric outcome. The methods in which electrons can be knocked out of metals by irradiation with light. According to classical physics, the electron's energy knocked out should depend on the intensity of the light. Strangely enough, this is not the case. The energy of the electrons does not depend on the intensity but the frequency of the light. Einstein could explain that.

For this, back again to the quanta of Max Planck. The energy of each quantum depends on the frequency of the electromagnetic radiation. The higher the frequency, the greater the energy of the quantum. Einstein now assumed, in contrast to Planck, that the electromagnetic radiation itself consists of quanta. The interaction of a single quantum with a single electron on the metal surface causes this electron to be knocked out. The quantum releases its energy to the electron. Therefore, the energy of the electrons knocked out depends on the frequency of the incident light.

However, the skepticism was great at first. Because electromagnetic radiation would then have both a wave and a particle character, but another experiment also showed its particle character. This experiment was conducted with X-rays and electrons carried out by the American physicist Arthur Compton (1892—1962) in 1923. As already mentioned, X-rays are also electromagnetic radiation, but they have a much higher frequency than visible light. Therefore, the quanta of X-rays are very energetic. That's why they can set a particular form. But that makes them so threatening. Compton was adept at viewing that X-rays and electrons act relatedly to billiard balls when they meet. This again showed the bit character of the electromagnetic radiation. So, their twofold nature, the alleged "wave-particle dualism," was lastly acknowledged. By the way, it was Compton who introduced the term "photons" for the quanta of electromagnetic radiation.

What are photons? That is still unclear today. Under no circumstances should they be imagined as small spheres moving forward at the speed of light. Because the photons are not located in space, so they are never at a certain place. Here is a citation from Albert Einstein. Although it dates back to 1951, it also applies to today's situation. "Fifty years of hard thinking have not brought me any closer to the answer to the question "What are light quanta? Today, every Tom, Dick, and Harry imagines they know. But they're wrong." .

The Bohr Atomic Model

We take atoms for granted. Their existence was still controversial until the beginning of the 20th century. But already in the 5th century BC, the ancient Greeks, especially Leukipp and his pupil Democritus, spoke of atoms. They thought the matter was made up of tiny, indivisible units. They called these atoms (ancient Greek "átomos" = indivisible). In his miracle year 1905, Albert Einstein not only presented the special theory of relativity and solved the mystery of the photoelectric effect, but he was also able to explain the Brownian motion. In 1827 the Scottish botanist and physician Robert Brown (1773—1858) discovered that dust particles only visible under the microscope make jerky movements in the water. Einstein was able to explain this by the fact that much smaller particles, which are not visible even under the microscope, collide in huge numbers with the dust particles. This is subject to random fluctuations. The latter leads to jerky movements. The invisible particles must be molecules. Therefore, the explanation of the Brownian movement was regarded as their validation and thus also as the validation of the atoms.

In 1897, the British physicist Joseph John Thomson (1856—1940) discovered electrons as a component of atoms and developed the first atomic model, the so-called raisin cake model. Therefore, the atoms consist of an evenly distributed positively charged mass in which the negatively charged electrons are embedded like raisins in a cake batter. This was falsified in 1910 by the New Zealand physicist Ernest Rutherford (1871—1937). With his experiments at the University of Manchester, he was able to show that the atoms are almost empty. They involve a small, positively charged nucleus. Around him are the electrons. They should rotate nearby the nucleus as the planets rotate near the sun. Another form of movement was inconceivable at that time. That led physics into a deep crisis. Because the electrons have an electrical charge and a circular motion causes them to release energy in electromagnetic radiation. Therefore, the electrons should fall into the nucleus. Hence the deep crisis because there should be no atoms at all.

In 1913 a young colleague of Ernest Rutherford, the Danish physicist Niels Bohr (1885—1962), tried to explain the atom's stability. He transferred the idea of quanta to the orbits of electrons in atoms. This means that there are no random orbits around the nucleus for the electrons but that only certain orbits are allowed. Each has a certain energy. Bohr assumed that these permitted orbits were stable because the electrons on them do not emit electromagnetic radiation. Without, however, being able to explain why this should be the case.

Nevertheless, his atomic model was initially quite successful because it could explain the so-called Balmer formula. It has been identified for some time that atoms only absorb light at certain frequencies. They are called spectral lines. In 1885, the Swiss mathematician and physicist Johann Jakob Balmer (1825—1898) found a formula with which the spectral line's frequencies could be described correctly. But he couldn't explain them. Bohr then succeeded with his atomic model, at least for the hydrogen atom. This is because photons can be excited by photons, causing them to jump on orbits with higher energy. This is the famous quantum leap, the smallest possible leap ever. Since only certain orbits are allowed in the Bohr atomic model, the energy and the frequency of the exciting photons must correspond exactly to the energy difference between the initial orbit and the excited orbit. This explained the Balmer formula. But Bohr's atomic model quickly reached its limits because it only worked for the hydrogen atom. The German physicist Arnold Sommerfeld (1868—1951) expanded it. However, it still represented a rather unconvincing mixture of classical physics and quantum aspects. Besides, it still could not explain why certain orbits of the electrons should be stable.

Sommerfeld had a young assistant, Werner Heisenberg (1901—1976), who, in his doctoral thesis, dealt with the Bohr atom model extended by Sommerfeld. Of course, he wanted to improve it. In 1924 Heisenberg became assistant to Max Born (1882—1970) in Göttingen. The breakthrough came a short time, in 1925 on the island of Helgoland, where he cured his hay fever. He explained the frequencies of the spectral lines, including their intensities, using the so-called matrices. He published his theory in 1925 and then his boss Max Born and Pascual Jordan (1902—1980). This is considered to be the first quantum theory and is called matrix mechanics. I will not explain it in more detail because it's not very clear. And because there is an alternative mathematically equivalent to it. It enjoys much greater acceptance because it is easier to handle. It is called wave mechanics and was developed in 1926, just one year after matrix mechanics, by the Austrian physicist Erwin Schrödinger (1887—1961).

Chapter 4.   The Heisenberg Uncertainty Principle

We've all been uncertain at times. Do we want the chicken or the fish? Which movie do we want to see? Eventually, you make a decision, and the uncertainty is gone. But to understand the next concept we're about to tackle, you need to think about being both certain and uncertain at the same time. Heisenberg's Uncertainty Principle, which he introduced to the world in 1927, aims to explain one of quantum mechanic's biggest problems—how can one predict where a particle will be at any given time, even with the knowledge of its momentum or prior position? First, let's take a look at Heisenberg's work that led him up to the Uncertainty Principle.

Heisenberg's Beginnings in Physics

Werner Heisenberg was born in Germany to academic parents. His father was a professor of ancient languages and Greek philosophy, and young Werner loved to engage in philosophical discussions with his teachers and peers. He spoke almost lovingly of the atom as a philosophical pursuit, which could only be reliably accounted for with mathematics. He would study under and with some of the other great scientific minds of his time, including Niels Bohr himself.

Heisenberg was also musically talented, a common thread among many of the pioneering physicists. His inclination for the piano guided him to meet his future wife, Elizabeth, after a performance. She was also from an academic family and encouraged him throughout his career to pushing his theories and research to new heights of discovery. The physicist was also an avid outdoorsman, active in many roles with the German Scouts throughout his lifetime. He would often retreat to the mountains when he was thinking through an incredibly difficult physics or mathematical problem.

While he is mainly known today for his famous uncertainty principle, Heisenberg's earliest major work was a collaboration born from his doctoral thesis. In partnership with Max Born and Pascual Jordan, Heisenberg proposed a set of mathematical matrices that could be used to describe and predict the motion of atomic particles to mechanical processes. Unfortunately for Heisenberg and his colleagues, they were in the Bohr camp of theoretical physics, which was slowly being phased out for the more progressive work of Einstein, Planck, Schrödinger, and de Broglie. While classical physics and mathematics were still a foundation of the newer fields of quantum physics, quantum mechanics, and atomic studies, the disciplines were experiencing a rapidly widening gap in beliefs and principles.

While Heisenberg's mechanical matrices were not universally accepted or used by the physics community, they weren't without merit. Part of the reason they fell by the wayside is that Bohr's school was falling out of favor as being outdated. While this seems a little ridiculous given the speed at which new quantum discoveries were being made, Bohr and his contemporaries and students were firmly entrenched in the physical properties of the atom as a real, tangible object. While the Einstein camp was studying wave-particle duality, the Bohr camp was concerned with what they called discrete bundles—quantum particles traveling together in packets of energy. They weren't interested in anything they couldn't measure through observation or predict with one hundred percent certainty. While Heisenberg would move away from his former colleagues in thought and action, due partly to Jordan leaving academia to become a Nazi SS officer in the 1930s, he would, later in life, give credit to Born and Jordan as being instrumental to his early development and eventual reception of the Nobel Prize. Heisenberg himself would spend much of the 1930s and 1940s under the scrutiny of the Nazis. They deemed his work to be counterproductive to their interest in harnessing nuclear power solely for weaponization.

The Development of the Uncertainty Principle

The Heisenberg Uncertainty Principle has become Heisenberg's lasting legacy in the world of particle physics, and it was a long time in development and refinement. Heisenberg would never completely leave behind his belief in the Bohr school of study. Still, he would eventually have to recognize that the work of those in the Einstein school was garnering more attention. Heisenberg's views of the studies conducted by those who were working in conjunction with Einstein were complicated. He saw their work as dealing in "reality" and considered himself an "anti-realist." The contradiction here is that Heisenberg loved the mathematics of physics, which deals mostly in, yes, reality. Numbers are absolutes and are very real. So, where did the Uncertainty Principle stem from?

Let's look at the basic premise of the Uncertainty Principle (which, by the way, Heisenberg himself called the Indeterminacy Principle). It states that it is impossible to know both the position and the momentum of a particle at the same time, even using observation, predictors, and equations. That's a pretty bold statement, so let's look at why it is both true and controversial. If you can see and measure something, then surely it is exactly what you think it is and where you expect it to be. To this day, scientists argue this point. Many feel that the measuring of particles with precision is the only way to be sure of their behavior. Why would Heisenberg be so uncertain about this?

Heisenberg postulated that the nature of the quantum movement is that there is a limit to how much knowledge one can gain from it. He believed that there were forces at work within a quantum system that were beyond the scope of human observation and understanding. Heisenberg went so far as to theorize that the more accurately one variable within a system could be measured, the greater the inaccuracy of another measurement. In simple terms, the more precise the measurement of the position of a particle, the less precise the measurement of its momentum, and vice versa. Why would he think this? And more to the point, could he prove it with his beloved mathematics?

Using his earlier presented mechanical matrices, Heisenberg set out to prove his indeterminacy principle, and sure enough, given the tiniest variations in particle movement and momentum, he proceeded to prove that at times b didn't always equal b time a. The infinitesimal differences he observed in quantum movement served as the mathematical basis for what would become the Uncertainty Principle. Remember, Heisenberg and his colleagues were concerned mostly with mechanical systems, meaning that the particles were not existing in a vacuum, as with electromagnetic systems. Heisenberg concluded that the existence of even the most minuscule outside forces was causing the atoms to behave in a way that made observation and measurement limited in scope, thereby limiting the knowledge one could gain from studying the system.

Because he was also a man of philosophy and action, Heisenberg also conducted what his cohorts deemed a "thought experiment," although Niels Bohr would later admit that the scientific basis of the research was sound. To perform this experiment, Heisenberg tried to examine the behavior of atomic particles, namely electrons, using a gamma-ray microscope. While observing these particles, he noticed that the gamma radiation was acting against the natural movement of the particles. It was essentially "kicking" the electrons around, not allowing him to get an accurate picture of what the particle should be doing in their natural state.

Heisenberg ultimately posited that it was a limitation of the nature of the quantum movement itself and not the scope or limitations of the observational equipment itself that created the uncertainty paradox. Every time he applied an observation tool that emitted more energy, he injected that energy into the system and further increased the uncertainty. In the most basic of terms, it is impossible to know what a particle will do, even if you know where it is. If you know what it's doing, you can't pinpoint its exact location. This principle is now one of the foundations of particle physics, quantum mechanics, quantum chemistry, and theoretical physics.

When scientists want to use Heisenberg's Uncertainty Principle in their work, they look at all the mitigating factors that could affect their measurements and observations, including the abilities and limitations of their laboratory equipment. They also consider the accuracy of their baseline data, the confidence they have in their prior or preparatory work and the work of others, and the earlier known uncertainty of similar experiments or materials. By gathering and collating this data before beginning an experiment, physicists and chemists can determine the potential for variations and margins of error within their research.

Mathematics, Uncertainty, and the Planck Constant in Action

When developing an equation to express the Uncertainty Principle in actionable terms, Heisenberg needed to employ Planck's Reduced Constant, or the h-bar. The simplest form of this equation is shown here:

$$\Delta x \, \Delta p_x \geq \frac{\hbar}{2}$$

This equation is a visual representation of the principle, and you can see Planck's Reduced Constant on the right side of the mathematical sentence. It is divided by two because there are two variables on the left side of the equation. On that left side, we see two Greek deltas, which are both uncertainties. The delta followed by the variable x represents the measurement of the position of a quantum particle, and the delta followed by the variable px, which represents the measurement of the particle's momentum. The delta itself stands for the standard deviation. When we put it all together, the entire equation reads, "The standard deviation of the position times the standard deviation of the momentum is greater than or equal to half of Planck's Reduced Constant."

Broken down like this, it's not hard to see what Heisenberg was getting at with his principle. The standard deviation is the amount above or below a predicted measured location where the particle can be expected to be found or its predicted measured momentum. This will vary by particle and condition, of course. He predicted that these amounts multiplied by each other would always come out to an equal or larger number than the reduced constant divided by the number of variables. If the number were ever to come out less, that would mean that the position and momentum of the particles were predicted with one hundred percent accuracy before they even came into the field of operation, which is statistically highly, highly improbable.

Chapter 5.   Photoelectric Principle Applications

We have seen how the theory of blackbody radiation fits the experimental evidence only if we assume that energy is always absorbed and emitted "packet wise," in the form of discrete quanta. However, blackbody radiation was only the first hint of this kind. Many other experiments showed how nature exchanges energy only in the form of quanta. These experiments told us the same story in different ways, but always with the same underlying plot, according to which energy and, therefore, also light can be conceived of as made of what we imagine to be particles. One of the most important constructs of evidence of this type came from the so-called "photoelectric effect," which conventional physics considers proof of the particle of light, the "photon" (a word coined by the American physicist's Gilbert N. Lewis). The photoelectric effect had already been discovered in 1887 by the German physicist H.R. Hertz. He observed that sending the light of sufficiently high frequency onto a metal plate leads to the emission of charged particles from the metal surface. These turned out to be electrons. Hertz, however, was not able to explain the true meaning and physical implications of the phenomenon. The theoretical understanding and explanation came in 1905, developed by Einstein. We will outline only a brief sketch here without going much into the technical details to identify what this is all about.

The aim is simply to give you an intuitive idea. Imagine having a metal plate onto which you shine a beam of light. Suppose the light is in the far-infrared spectrum, much below the red-colored light, which means we are using only a relatively low-frequency (long wavelength) EM wave. This low-frequency light beam would have no effect on the metal plate (except, of course, that it would heat up or eventually melt); no emission of charged particles, the electrons, would be observed as long as the light was under a specific threshold frequency (or, equivalently, above a particular wavelength). This occurs regardless of the beam's intensity; it will not happen even for strong light intensities. If the light goes beyond a threshold frequency, suddenly, a flux of electrons from the metal surface can be detected. Again, this occurs regardless of the beam's intensity; it also happens for low-intensity light. This shows that light waves can extract the electrons, the so-called "photoelectrons," from the atomic metal lattice. This occurs only for individual metal plates, making it clear that the photoelectric effect does not extract electrons directly from the atoms. Photoelectrons are trapped inside the atomic lattice and are nevertheless free to move inside when an electric field is applied. This property makes good conductors out of certain metals and allows us to transport electric energy through metal cables.

Therefore, this extraction process from the metal lattice only occurs when the electrons acquire specific energy at a minimum extraction frequency (or extraction wavelength). If we trace along the horizontal axis the frequency of the incident light and along the vertical axis the kinetic energy of the outgoing photoelectron (not to confuse with the momentum), we can see how only if the wavelength becomes small enough (or, equivalently, only if the frequency becomes high enough, that is, larger than), the electrons will start being "kicked" out from the surface and appear with some kinetic energy larger than zero. This implies that part of the light life incident on the metal surface must be employed to extract the electrons first from the metal lattice. This minimum energy is called the "work function." Every metal has a different work function, which means that every metal has a different threshold frequency. However, the effect remains qualitatively the same. Once the electrons are extracted from the plate, what remains of the energy of light will go into the electron's kinetic energy. So, in general, it turns out that the observed electron's kinetic energy must be the difference between the power contained in the incident light photons and the amount of work necessary to extract them from the metal. Note that this is not theoretical speculation but an exclusively experimental fact; measuring for each light frequency with which one shines the metal plate and the kinetic energy of the electrons for that frequency, one can plot it into a graph and obtain a strictly linear dependency between the light's incident energy and the emerging electron's kinetic energy. While the threshold wavelength (or threshold frequency) is determined by the type of metal one uses. Every metal and kind of substance one uses has a difference—that is, the straight-line function is only shifted parallelly upward or downward but does not alter its form or steepness. The straight line of the slope is always identical. The slope of the kinetic energy linear function is a universal constant, namely, Planck's constant. This is how Planck's constant could be measured precisely. Moreover, increasing the intensity, which increases the number of incident photons, does not change the result. Higher light intensity does not lead to more energetic ejected electrons; it only determines its number. This is somewhat unusual for a classical understanding of where we imagine light as a wave. Why does light extract electrons from the metal only above a threshold frequency (say, in the spectral domain of ultraviolet light or x-rays) but not below it? Why can we not bombard the metal surface with a higher-intensity light of the same 40 wavelengths (imagine many waves with large amplitudes) to obtain the same result? How should we interpret this result?

The generally accepted explanation, according to Einstein, is that, as we have seen with the blackbody radiation, we must conceive of light as made of single particles—that is, photons carrying a specific amount of energy and impulse, and which are absorbed by the electrons one by one. However, this happens only if the single-photon has sufficient power to extract the electron from its crystal lattice in the metal. The electron will not absorb two or more photons to overcome the lattice's energy barrier in which it is trapped. It must wait for the photon with all the necessary energy that will allow it to be removed from the metal lattice. And because it absorbs only one photon at a time, all the electrons emerging from the plate will never show up with energies higher than that of the single absorbed photon, even though a larger number of photons (higher intensity of light) can extract a larger number of electrons. Einstein issued his theory of the photoelectric effect alongside his historic paper on SR. For this explanation of the photoelectric effect and his "corpuscular" (little particles) interpretation of light, Einstein received his Nobel Prize—not for relativity theory, as is sometimes wrongly believed.

The Laws That Govern the Probabilistic Nature of the Quantum World

It is an easy experiment that you can do by yourself—just take a piece of paper, poke a little round hole with a pin or needle, and look towards a light source. You will observe the light source's image surrounded by several concentric colored fringes (the colors appear only because, fortunately, the world we live in is not monochromatic). The diffraction that occurs whenever a wave encounters an object or a slit, especially when its size is the same as the wavelength, the plane wavefront is converted to a spherical or a distorted wavefront, which then travels towards the detector screen.

In this case, you see that, even though we are dealing with only one slit, some weak but still clearly visible secondary minima and maxima fringes can be observed. True, it is easier to produce more pronounced interference patterns with more than one slit (or pinhole). For most applications, especially when the wavelength of the incident wave is much smaller than the aperture's size, these effects can be neglected—transverse plane wave incident on a single slit with aperture. However, strictly speaking, a single slit also produces small diffraction and interference phenomena.

Chapter 6.   Schrodinger's Quantum Theory

One of the most famous names in quantum physics is Erwin Schrödinger, although it's likely you've heard of him more for his thought experiments than his actual science. Schrödinger was an Austrian-born scientist who emigrated to Ireland and finished his career after a little run-in with the Nazis leading up to World War II. He's also famous for his "feud" with Einstein due to Schrödinger's so-called failing to embrace the fundamentals of quantum physics as early as some of his colleagues. We'll talk about his infamous cats and his philosophical pursuits in a little while, but first, let's see what he did in the laboratory over his storied career.

Schrödinger's Quantum Theory and Unified Field Theory

In the 1910s, Schrödinger became acquainted with the works of Planck, Einstein, and others but wasn't interested in giving up classical physics just yet. That all changed when he developed tuberculosis and spent some time in and out of sanatoriums, regaining his health. While convalescing, the scientist started to warm up to the ideas placed before him, especially quantum theory. He was fascinated with the possibility that the electromagnetic spectrum could have differing effects on a wide variety of elements and wanted to discover if he could find a universal way to predict the behavior of electrons.

Schrödinger played around with these thoughts and concluded that the only way to do this would be to be able to predict the nature of radiation itself. While he couldn't find a way to do this, the articles he published proposing these concepts opened the door for a new brand of theoretical physics and did lead to a few concrete advancements, such as wave mechanics and an interesting new atomic model, which we'll get into shortly. Schrödinger's other contribution to theoretical physics was his attempt to create a unified field theory, which is sort of the Holy Grail of quantum physics. A unified field theory, or UFT, had been attempted by Einstein when he was working on his theory of relativity and is still a matter of great debate among quantum scientists.

A unified field theory would bring together all the fundamental interactions and relationships in quantum physics under one set of proven laws. This would mean that researchers would be able to predict the behavior of all matter based on mathematical proofs and observable actions with no deviations. The matter would behave according to these laws, and we'd be able to connect the dots between interactions, electromagnetic fields and waves, particles, and space-time. There have been more than a few attempts to write and prove a unified field theory for quantum physics (but classical physics has one), but no one is yet to succeed. One such popular attempt is called "the theory of everything," and let's face it, that does sound a bit pretentious.

Wave Mechanics

One thing that Schrödinger was successful in was becoming the father of wave mechanics. His theories on the behaviors of particle waves gave birth to this new sub-discipline in the mid-1920s. The basis of his theories was rooted in the premise of the behavior of a hydrogen atom in a system independent of time constraints; Schrödinger wanted to know what would happen if time was taken out of the equation when predicting the wave-particle behavior of the atom. In the resulting rapid-fire series of four papers, the physicist laid out his predictions and gave the world the first look at his now-famous equation; the second paper gave the equation an edit to account for harmonics within the system.

The third and fourth papers in the series were devoted to showing the world how to compare his equation to work involved in the uncertainty principle and taught his colleagues how to plug in the proper complex numbers directly into his equation to avoid having to calculate so many derivatives. This series of papers is still considered one of the greatest scientific accomplishments in modern science. It paved the way for physicists to begin studying wave behavior in a much more controlled and accurate manner.

It also marked the beginning of some seriously complicated math, and many believe Schrödinger's equation to be the cut-off between classical and quantum physics. Do you see now why it's so difficult to pinpoint one defining moment? Even Schrödinger himself wasn't thrilled with what he'd done once it was over and published. He recognized that he had caused a giant rift between classical physics and the new quantum discipline. As someone who loved the principles of classical physics, he never wanted to be as right as he was- but once his research was published, there was no going back.

Schrödinger's Equation and Atomic Model

What exactly was so special about Schrödinger's equation, other than the fancy numbers? And why did it have such a profound effect on quantum physics? To begin with, it is a partial linear differential equation, which means it has a lot of moving parts to get to its solution. Heck, that's a lot of words just to describe an equation. In classical physics, there is Newton's second law of motion, which you may recall is shown mathematically as F=ma or force equals mass times acceleration. It predicts the movement of an object as it speeds up, proportional to its mass. Think of Schrödinger's equation as the quantum physics counterpart to Newton's second law. Yes, it was that groundbreaking and that important. Below here, you will see the equation:

$$H(t)|\psi(t)\rangle = i\hbar\frac{\partial}{\partial t}|\psi(t)\rangle$$

Yeah, we know. This is a lot to unpack, so we're just going to pay the baggage fees on this one and send it on its way. After all, this is a book for beginners. This sucker took F=ma and turned it on its classical physic's head. But for those who are capable of looking at Schrödinger's equation and doing the math, it is crucial. Without this equation, scientists might still struggle to predict the behavior of wave functions over time. With the equation, a little simple math gives them all the answers they need. It's said that with this one calculation, Schrödinger flung open the door of quantum mechanics, and it was never able to be shut again.

Schrödinger was also intrigued by the thought of updating the atomic model to account for the wave behavior of electrons as they orbited the nucleus. After careful analysis, the scientist was able to create the first truly three-dimensional, accurate model of the atom, using his wave mechanics theory and equation. Schrödinger's model of the atom includes an electron cloud, moving in and out around the nucleus in a wave pattern. Using this model, scientists can predict where the electrons might be at any given time. This differed from the earlier Bohr model that showed the electrons in a set, layered orbits that did not fluctuate due to wave-particle duality.

Chapter 7.   Schrodinger's Cat Paradox

When talking about entangled particles and superimposed states of the same particle, reference is often and willingly made to Schrödinger's cat paradox experiment. The great physicist was able to describe a mental experiment (don't worry for the kitten, he never really performed it!) that transferred the superimposition of two states of a particle (in this case, an atom) to a macroscopic object, that is a cat. In this way, he confronted his colleagues with a verifiable absurdity in everyday reality.

In reality, Schrödinger had proposed it only to provoke and demonstrate how extending the concept of "overlapping states" required a lot of attention and led to unacceptable solutions. In other words, he who had derived the wave function, the basis of the whole QM, had many doubts about a completely probabilistic and not deterministic view of the world (he was, in short, on Einstein's side).

Often, on the other hand, his "cat" is presented as an experiment that proves just the opposite of what the author himself thought. Common problems in the divulgation of what is not well understood by those who divulge.

It is worth, however, describing it and using it to summarize many ideas and try to understand how it can be "disassembled" in various ways without violating the basic rules of QM.

Let's take the exact words of the Austrian physicist, which make his motivations clear to us:

"One can even set up quite ridiculous cases. A cat is penned up in a steel chamber, along with the following device (which must be secured against direct interference by the cat), in a Geiger counter, there is a tiny bit of radioactive substance, so small, that perhaps in the hour one of the atoms decays, but also, with equal probability, perhaps none; if it happens, the counter tube discharges and through a relay releases a hammer that shatters a small flask of hydrocyanic acid. If one has left this entire system to itself for an hour, one would say that the cat still lives if, meanwhile, no atom has decayed. The psi-function of the entire system would express this by having in it the living and dead cat (pardon the expression) mixed or smeared out in equal parts."

It is typical of these cases that an indeterminacy originally restricted to the atomic domain becomes transformed into macroscopic indeterminacy, which can then be resolved by direct observation. That prevents us from so naively accepting as valid a "blurred model" for representing reality. In itself, it would not embody anything unclear or contradictory. There is a difference between a shaky or out-of-focus photograph and a snapshot of clouds and fog banks."

Before I continue, let me make a small and insignificant modification since I am a gutted lover of cats and animals in general. I would prefer to replace the flask of hydrocyanic acid with a nice anesthetic (rightly dosed) that makes the cat safely sleep deeply for at least an hour. All right? Ok, let's continue.

The case of the cat is easily attackable and rebuttable. Various ways have been used. For example, we can say that the superimposition of state can happen only in a closed system (somehow as if the chamber represented a Universe separated from ours), while instead, this is not true and through any kind of radiation or very thin interrelations between the chamber and external world this does not happen, and therefore the wave function collapses immediately, and the cat is really either alive or dead even if we cannot "apparently" ascertain it.

Or, we can say that the cat is never in the superimposition of states because it is already broken by the Geiger counter, placed between the radioactive atom and the poison, which serves precisely to release the poison. This is, to all intents and purposes, an observation.

Let's not forget, however, that still today, we think of a much more fascinating solution, the cat is really in two superimposed layers that both exist in two separate universes. In one, the cat is alive, and in one, the cat is dead.

In short, the implications and discussions, not wanted by Schrödinger himself, continue today and lead to truly extraordinary visions and theories even if purely theoretical and not verifiable with "ad hoc" experiments. We remember that the oddities of the microcosm are now perfectly verifiable (as we have already seen partially) and used in the most advanced technology in all fields of science. But not only that. Phenomena like the nuclear fusion inside the sun star (and its numerous sisters) can happen only and only if certain principles of the QM occur. In particular, the tunnel effect, which is linked to Heisenberg's uncertainty principle, really commands all "microscopic" dances.

There remains, however, a profound doubt still unresolved. What is it, and where and how is the boundary between the microscopic and macroscopic world established? The application of the uncertainty principle succeeds in resolving much of this problem. Even stronger is the doubt about the actual need to have a conscious verification of the position of the particle. Is the Geiger counter (devoid of intelligence) enough to transform the wave into a particle and/or a device that does not warn us in any way from which slit the photon has passed or is it really necessary an effective awareness of the phenomenon? On this still open question, we will return shortly in the conclusions.

Let us not forget, finally, that the non-deterministic view of reality is far from resolved. The problem of missing variables raised by Einstein and colleagues seems to regain vigor, and a recent macroscopic experiment seems to open new visions of reality, bringing us back to de Broglie's famous pilot wave. A kind of deterministic compromise in which all the descriptions of the QM would remain unchanged but would be controlled by a kind of material wave that guides the particles, which would never actually acquire the "state" of waves but would adapt to the pilot wave. A very preliminary example? Think you have a liquid that has no evident vibrations but that is traveled by waves below the detection threshold. Drop a drop of the same liquid. You'll see it disintegrate and move into the liquid pushed only by the wave it caused. The drop remains a drop but takes on the characteristics of a wave.

Before moving on to the preliminary conclusions (remember that this is only a guide for beginners), which summarize and generalize what we have seen so far, let me recommend to the most willing a mind game that requires a great logical and interpretative effort. Remember the famous board games in which situations such as "One door leads to Hell and one to Paradise. In front of each door, there is a guardian; one always tells the truth, and the other, a lie. We do not know, of course, which of the two is the liar and which is the honest one. What question do we have to ask either of the two guardians to make sure we go to heaven?"

Chapter 8.   Wave Theory

These are two usual aspects of Einstein's field practice and Maxwell's electric field. One more approach to take a look at the quantization cycle is to at first overhaul field conditions corresponding to mathematical administrators who consolidate those numerical standards

There are a large number of establishments for the use of quantum field speculation. In any case, a typical hypothesis of traditional style convictions, which is one of our best (non-esteem) things for the contemplation of nature. Second, quantum field theory can speak to (perceptions, adjusted suspicions) the creation and crumbling of particles and non-logical proportions of a quantum material. Third, the quantum field theory is relativistic inherently and "mysteriously" handles complex issues that plague even the quantized atom hypotheses.

Be that as it may, no, quantum fields are not viable with any kind of effect. Quantum fields are significant. In quantum field speculation, what we see as particles is a fascinating field of the quantum field.

Quantum electromagnetism is an unmistakable "convenient" theory of the quantum field. There are two fields in it, the electromagnetic field and the electron field. Along these lines, for instance, what we normally observe as an electron-focused electron is a sure association in quantum electrodynamics between an electromagnetic field and an electron field, where the electromagnetic field loses quantum incitement, and the electron field assimilates its strength and performance power.

How would you explain the idea of the quantum wave of the problem?

What Is a Wave

- It generally quantifies the range of a sine wave, i.e., the distinction between the broken-down waveforms.
- Frequency measures how often the sine wave is reset in a second.
- Quantity quantifies the size of the frequency scale over zero levels.
- The stage decides the situation of the point on the wave in the second situation in the space, in recurrence units.

Wavelength Measurement

- Size A
- Wavelength λ
- Category Shift $\Delta\varphi$

Interruption

It is extremely useful to use wave impedance to discover superfluous allotments. If two wave surfaces are suspended, their non-wave pinnacles may abbreviate (gainful impedance) while confronting a higher value, and the body, by and large, will radiate a wave. The example of ruinous and ensuing impedance in space makes it simple to envision recurrence.

It is an element of material science that the lifting power is not identified with the dense concentrations of particles and the fragile set of a single atom in a machine at some random time. The capacity of the wave, despite everything, demonstrates a true quantum object. This is one motivation behind why, sometimes, it implies that "all cells are isolated."

Quantum theory can just recognize the likelihood of a specific result. Which of these expectations is finally expected in the essential inclusion of the overall political race and the regions? Just a couple of appraisals under similar conditions show cautious dissemination of chances, which is additionally taken out from Schrödinger's framework.

$$i\hbar\, \partial/\partial t\, \psi\,(r, t) = (-\hbar2/2m\, \Delta + V\,(r, t)\, \psi\,(r, t)$$

All designs to date have demonstrated the following, square modulus | ψ | 2 of the state work ψ refers to the probability of getting a quantum object during special position t and all the various boundaries contained in ψ.

Slender Film Disruption

The obvious impacts of impedance are not restricted to the twofold-edged computations used by Thomas Young. The impact of a small block of the film is caused by the light that shows the two zones isolated by a range equivalent to their size.

The "film" in a space can be water, air, or some other indistinguishable or strong fluid. In splendid light, the obvious impedance impacts are restricted to films with the extent of a couple of microns. The notable model is an air pocket cleaner film. The light reflected from the air pocket is a two-wave lift, one obvious on the front surface and the other on the back. Two waves show spread and interruption into space. The size of the cleaning film decides if these two waves can meddle with help or in a dangerous manner. The full test shows that considering only the recurrence λ; there is a useful film thickness impedance equal to λ/4, 3λ/4, 5λ/4, and destructive interference for thickness 3λ/2.

As the white light enlightens the cleaning film, the hidden gatherings are viewed as different frequencies moving through destructive obstacles and isolated from the show. The reflected light is consistently shown as a comparing shade of the irradiated recurrence (e.g., when a red light is produced with a damaging impedance, the splendid light shows up as cyan). Thin-oil films produce a near impact on water. In Nature, the quills of winged creatures, including peacocks and fowls, just as the shells of specific creepy crawlies mirror light as the shade of the significant changes with a modified state.

This is achieved by restricting intelligent light waves from misleadingly planned structures or by the wide variety of show posts. Subsequently, abalone pearls and shells sparkle from the restriction brought by presentations from different pieces of mother-of-pearl. Stones, for instance, opal, show the flickering impacts of gleaming that originate from dissipating light from the commonplace instances of round particles.

There are numerous uses of hindering light impact mechanical impacts. Inclusion's main enemy is the camera's center focal point that focuses on little estimated films and recovery records taken to make the impedance of a risky showcase of clear light. Constant progressed inclusion, which incorporates different slight film layers, is made to transmit light only within a narrow range of wavelengths and thus fill as recurrence channels. Multilayer textures are also used to enhance the mirror presence on infinite telescopes and laser optical gaps. Genuine interferometry methodology measures small changes in related isolation by observing turning shifts in light-hindering plans. For instance, the state of Earth in the obvious parts is reflected in the optical waves segments using interferometry methods.

Wave-Fields and Interference

Like in the whole nature, it is conceivable that understanding has developed dynamically, and we think that for us people, it is over. However, we are not a gathering with a decent comical inclination. But this can dramatically decrease; it is restricted to living creatures with neurons. Few can guarantee that cell life alone is conceivable or that plants have it (and I likewise incorporate reptiles).

Regardless, we are powerless as far as where we are gaining little ground, which is 1.5 billion after the primary acknowledgment of Great Climacteric. Neurophysiology, with all the neuroscience involved, will discover nothing like that.

Other than that, there is a favorable position in our stockpile that we, regardless of all, have not yet used; the information loupe. At its most honed point, a wonderful spot in the region was found a few million years back by the Great Climacteric sign. We will see where evolution switched gears in its neuronal endeavors and began another cycle.

Cognizance Polychrome

Since the mid-1960s, tasks have been progressing at NASA and at different institutions for the proof of intelligent life in space—SETI exercises. If these endeavors pay off, we may receive or find an answer to our messages. One such message will tell us there is life on the farthest planet and where it is.

Chapter 9.   Electromagnetic Theory

All we know consists of matter made up of particles, that is, atoms, which consist of smaller particles, electrons, protons, and neutrons (the last two are made up of quark particles), and invisible energy interactions or forces, among which are gravity, the attractive and repulsive forces of charged objects or magnetism, X-rays, gamma rays and radio waves (mobile phones, Wi-Fi, radio and TV).

There are only four invisible forces. The first one, gravity, is vital for our environment as it ensures that all objects around us stay in place, including keeping each of us on the ground as well as keeping the Earth in a consistent orbit around the Sun. Yet, it is the least strong force of all. There are two much stronger forces inside atomic nuclei (called simply weak and strong interactions), which ensure the existence and stability of the nucleus of each atom in the known universe (including the atoms that make up each of us). The rest of invisible reality (outside gravity and nuclear forces) is classified into the subtypes of electromagnetic forces or electromagnetic radiations (each force always has a source, that is, it is always radiated by something), including the above-mentioned radio waves, X-rays, and gamma rays, as well as some others, as we will see.

Now, to begin the study of quantum physics, it is necessary to have a general understanding of electromagnetic radiation, as quantum physics was born out of attempts to gain an insight into it.

It is easy to understand the quantum hall function using the behavior of elements in a magnetic field. You can understand how physical processes work in cases of solid-state physics and electrical charges. Honestly, you need to know how electrons react in a magnetic field.

Research conducted by a team of professionals used a transmission electron microscope to produce nanometer-sized vortex electron beams. There were various quantum angular momentum states, and they evaluated the beam circulation to reconstruct the rotational dynamics of the elements in different states.

The electrons involved in this study are supposed to rotate easily with cyclotron frequency. This is the incidence applied by a charged element traveling through a magnetic area. However, the team revealed that, depending on the quantum integer expressing the angular momentum, the electrons rotated and moved in three separate directions using the frequency of zero. Also, the electrons could travel using Lamor's frequency and the cyclotron frequency, respectively.

With this experiment, it is proven that the rotational dynamics of the elements are difficult and complex, and different from human anticipations.

At first, we knew nothing about invisible forces. We could only say that in addition to matter, there were forms of pure energy, which caused various effects, such as heating, destruction, attraction, and repulsion. As all human thought, science, and technology developed, we learned more and more about them. People were yet to learn that they were studying electromagnetic radiation while researching its most accessible subtype, light. The double-slit experiment showed that light behaves like water or sound in that it moves in waves. All waves have the same features; wavelength, frequency, and amplitude. Waves can be imagined as a wavy line with alternating regular peaks and troughs. Wavelength is the distance between two neighboring peaks or two neighboring troughs (or the distance over which the wave's shape repeats). Frequency is the number of waves per second (the velocity of these waves is the light's velocity, namely, 300,000 kilometers per second. Thus, the frequency is the number of waves per 300,000 km of radiation). The smaller the wavelength is, the more waves there are within one second or per 300,000 km; that is, the more the frequency will increase. Thus, wavelength and frequency are directly interconnected; each wavelength value corresponds to its frequency value—inversely proportionally. Amplitude is the height of a wave peak or the depth of a wave trough (measuring from the median line). Its value is the energy value of waves.

Waves also have a curious and important property; when two or more waves meet, they overlap and interact in such a way that if the peaks of one wave meet the peaks of another, they are added together. That is, their amplitudes are added, and a combined higher peak is formed. The troughs of both waves are also added, forming deeper troughs, but if the peaks of one wave coincide with the troughs of another, the waves neutralize each other.

This means that water or lightwave energy is redistributed in space from the neutralization points to the points of new peaks where the earlier peaks crossed. As a result, at the points of neutralization light disappears, and intensifies at the points where the amplitude increases. As will be described later, the ability of the waves to interact this way makes it possible to distinguish between waves and particles in experiments.

Later on, other types of radiation, which scientists had begun to detect and study, were also found to be energy with wave properties. All types of radiation (except particle radiation) were found to be parts of a single spectrum of electromagnetic radiation, differing only in wavelength (or frequency). Radio waves (wavelengths from 1 decimeter and over, up to the size of the radiating object), microwaves (from 1 decimeter to 1 millimeter), infrared waves (1mm to about 1 micrometer, that is one-thousandth of a millimeter or 1,000 nanometers), regular light visible to the human eye ("white light," which is a uniform mixture of a range containing all colors, from red to violet) (750 to 400nm), ultraviolet (400 to 10nm), X-rays (10 to 0.01nm), and gamma rays in whose range the spectrum is infinite since the wavelength can decrease to ever smaller and smaller values, closer and closer to zero (0.01 => 0.001 => 0.0001 => ...), infinitely.

56

Electromagnetic Interactions

You can test the forecast of a nuclear model using the electromagnetic interactions of the nucleus of that model. Ground-level magnetic dipole and electric quadrupole, including the excited states, can help us properly understand the nuclear structure. You can guess that the operator of the magnetic instance of a state examines the authenticity of the wave effect.

Depending on the distance, the electromagnetic interface of energies can occur between polarized systems. In optical physics and atomic or molecular physics using Van der Waals, potentials are used that are produced by joining a pair of polarized systems such as an atom, a molecule, an electron, or ion, causing divisions in long ranges with negligible exchange forces.

Radial Wave Function

Here, we will learn the operations of the radial wave of a hydrogen molecule, a common atom with an electron and proton as constituent elements. The radial wave effect can help in determining the energy level and the possibility of discovering an element in a hydrogen atom. The function of the radial signal of an atom is reinforced by the principal quantum numeral (n), including the orbital quantum integer represented by l.

The Time-independent of the Schrödinger equation is applied in cases involving globular coordinates and variable separation methods. The radial wave of the hydrogen molecule could be normalized to produce some wave functions.

Chapter 10. Compton Effect

As with everything in quantum mechanics, equations are the math created to explain a variety of events on the molecular level. Scientists are always looking for a better way to explain how electrons as expressed by light or other matter are moving and the energy released as well as gained through that movement. One such equation was created by the Compton Effect, otherwise known as the Compton scattering. It was found to come from a high-energy photon engaging a certain target within a collision. Thus, the process allows the release of loosely bound electrons out of that outer shell found as part of the molecule or a specific atom.

As a result of the collision, scattered radiation practices a shift in the wavelength that didn't fit into the classical wave theory, remember, the classic wave theory has been taking a beating, so to speak, from these experiments and hypotheses so focused on how electrons and matter can be moving in terms of particles and waves. This is yet another blow to the classic wave theory. As we have seen with all these experiments, most of them start with the premise of Einstein's photon theory and appear to show support for that theory.

Arthur Holly Compton received a Nobel Prize in 1927, but the effect named after him was originally demonstrated in 1923. So how does this process known as the Compton Effect work? Simply put, the gamma or x-ray high-energy photon hits a defined target that has loosely bound electrons on the outer shell. This photon is known as the incident photon, is defined with the following energy E and linear momentum p. Within the Compton Effect, the photon gives a portion of its energy away to another almost free electron in the form of kinetic energy, which is to be expected when you have a particle collision.

Scientists have come to understand that energy and linear momentum must be preserved. When analyzing these relationships, three equations are the result. These equations include energy and x- and y-component momentum.

There are also four variables involved as listed below:

- Phi: An electron's angle of scattering
- Theta: Which is the photon's scattering angle
- Ee: Which is the electron's final energy
- E': Which is the photon's final energy

Suppose we only focus on the photon's direction and energy; then, we can treat the electrons as a constant. As a result, we can potentially solve the system of equations for effect. Compton combined several equations, and using a few tricks he picked up from algebra to eliminate some variables; he was able to create the two equations that are related because the energy and wavelength are both related in photons.

The Compton Wavelength of the Electron has a value of 2.426 x 10 -12 m. This value can be used as a proportionality constant designated for a wavelength shift. So why does this particular effect support protons?

In part, this analysis and derivation are based upon a particle perspective. The results have been easy to test. When observing the equation, the shift can easily be quantified in the angle's terms from which the photon is scattered. Simply put, everything on the right side of the equation is used as a constant. Since experiments have consistently shown this to be the case, thus supporting the photon interpretation of light.

Understanding some of these theories and the experiments behind them are important to have a greater understanding of quantum physics as a whole. However, nature always throws curve balls. So, it is no surprise that there are effects that cannot be explained through these theories. So how do scientists define the uncertainty inherent in this study of the smallest things on earth known as quantum mechanics? One such way is by the cornerstone of quantum physics, otherwise known as the Heisenberg Uncertainty Principle.

Chapter 11. Bohr's Hypothesis

With an idea of dazzling audacity, Bohr addressed both problems. Classical electrodynamics is straightforward; an electron circling the nucleus accelerates and would also emit radiation. It'd be similar to a small radio transmitter that broadcasts electromagnetic waves. This must waste energy in the process, sink further into the nucleus's enticing pull, and finally crash into the nucleus.

Bohr merely said that this wasn't the case. He stated; instead, there are stable orbits clustered around the atom's nucleus where an electron can orbit forever without losing energy.

Then, according to Bohr, electrons will jump between these permitted orbits. An electron must accumulate energy to hop away from the central nucleus and into a higher energy orbit. This allows it to transcend the atom's positive nucleus's attraction and travel away from all of this. Light dropping on the atom provides additional energy to it. The light's energy is passed to the electron, allowing it to leap to a high-energy orbit. The light should deliver precisely the correct quantity of energy to compensate for the energy gap between the orbitals.

Furthermore, Bohr believed that the volume of energy extracted from the thrilling light follows Planck's formula; energy equals h times frequency.

As a result, only light of a very particular frequency will excite the leaps between two orbitals. For a given leap, the light's frequency must be specifically calibrated such that frequency x h equals the energy required to achieve the leap. The reverse mechanism is also possible according to Bohr's hypothesis. An electron that has skipped to a higher orbit can't remain there indefinitely. It would descend to a lower orbital once more. It would then re-emit the energy it received from leaping up at a certain frequency of light.

The energy of the light produced would be equivalent to h x frequency, as predicted by Planck's formula.

As a consequence, when an electron leaps between two orbitals, it absorbs light with a certain frequency that is exclusive to that leap.

These amounts of light that it consumes and releases eventually became known as light quanta (Einstein). Bohr, on the other hand, was careful in 1913. He made no mention of a light quantum being emitted. He merely stated:

"The emitting of relatively homogenous radiation where the Planck's theory gives the relationship between the amount of energy transmitted and frequency."

It wasn't by chance that Einstein's quantum was avoided. Bohr was a staunch critic of Einstein's theory and only changed his mind after modern advances in the 1920s forced Einstein's theory to be accepted.

Bohr was able to deduce the strangest finding from the measured atomic spectra after making certain assumptions. As only a few light frequencies were available, only a few hops were feasible, resulting in only a few orbitals being allowed for the electron. It was as if our sun permitted a comet to orbit between Mars and Earth but barred every planet in between.

It was just a matter of determining which of the several alternative orbits are located in this favored set of stable orbitals. It was a pretty simple task. The differences of energy between these permitted, stable orbits were cataloged in detail by the observed spectra. Every line in the spectra was caused by electrons moving between two different orbits. It's a computational experiment to figure out which of the few orbits are which. The estimate was quite similar to this geography exercise. If we know the lengths between each couple of cities in a world, we can use that information to find out where each city is on the globe. The energy differences between Bohr's permitted orbits were determined by atomic spectra. He was able to calculate the energies and, therefore, the positions of the permitted orbits using these results.

When Bohr did this, he discovered a rather clear way to summarize the orbits. They were the ones whose angular momentum was expressed in h/2pi units. Bohr discovered that circling electrons always have to have whole units of angular momentum; two h/2pi, seven h/2pi, eleven h/2pi, with nothing in between, much like Planck's comparison tried to tell us that radiant energy is coming for whole units of frequency x h. We've shown that an object's ordinary momentum is equal to its mass multiplied by its velocity. Angular momentum is essential in the dynamics of spinning or orbital structures. It is known as an electron's orbital radius x angular speed x mass for a small particle like an electron circling a neutron.

Bohr's model was perplexing, to say the least. It appeared to demand that traditional physical concepts both retain and collapse at the same moment, much like Einstein's light quantum hypothesis. It was not a pleasant position to be in. A more positive reality overshadowed those annoyances. The hypothesis of Niels Bohr succeeded, and it operated well enough. Observational spectroscopy provided researchers with a large database of spectra from a variety of samples under a variety of conditions. Scientists were able to construct an enormous amount and fruitful account of them based on Bohr's theory. Although it was obvious that something wasn't quite right, it was appealing to put off these worries in the face of excellent results.

The angular momentum that was orbiting electrons possessed arrived in absolute multiples —quanta—in h/2pi, according to Bohr's theorem of 1913. In the years that followed, that simplistic condition was extended into a wider condition in which the quantity "action" was only available in entire multiples for physical processes that reverted to the same original condition periodically. As a consequence, physicists began to use the expression "quantum of action." This sidebar should have a single sentence that describes the physical quantity "action."

Chapter 12. Photoelectric Effect

When electromagnetic radiation of appropriate frequency is made to hit the surface of metal like, say, sodium, electrons are emitted from the metal. This phenomenon of emission of electrons from certain materials (which include several metals and semiconductors) by electromagnetic radiation is referred to as the photoelectric effect.

A metallic emitting electrode (E) and a collecting electrode (C) are enclosed in an evacuated chamber in which a window admits electromagnetic radiation of appropriate frequency to fall on E. A circuit made up of a source of EMF (S), a resistor (R), and a sensitive current-meter (A) is established between E and C. The polarity of S can be changed so that C can be either at a higher or a lower potential concerning E.

Features of Photoelectric Emission

This arrangement can be used to record several exciting features of photoelectric emission. If for a given intensity of the incident radiation, the potential (V) of C to E is positive, then all the electrons emitted from E are collected by C, and A records a current (I). This current remains almost constant when V is increased because all the photoelectrons are collected by C whenever V is flattering. This is known as the saturation current for the given intensity of the incident radiation.

However, this entire phenomenon of a current being recorded due to the emission of photoelectrons from E is dependent on the frequency (v) of the radiation. If the frequency is sufficiently low, then photoelectric emission does not occur, and no photo-current is recorded. For the time being, we accept that the frequency is high enough for photoelectric emission to take place. If holding the frequency and intensity of the radiation constant, one now reverses the polarity of S and records the photocurrent with an increasing magnitude of V. One finds that the photo-current persists but gradually decreases till it becomes zero for a value V = −Vs of the potential of C concerning E. The magnitude (Vs) of V for which the photocurrent becomes zero is termed the stopping potential for the given frequency of the incident radiation.

On the other hand, if the testing is repeated with different values of the frequency, keeping the intensity fixed, one finds that the stopping potential increases with frequency. One finds that if the frequency is made to decrease, the stopping potential reduces to zero at some finite value (say, v0) of the frequency. This value of the frequency (v0) is found to be a characteristic of the emitting material and is referred to as the threshold frequency of the latter. Indeed, no photoelectric emission from the material under consideration can take place unless the frequency of the incident radiation is higher than the threshold frequency. Moreover, for v > v0, photoelectric emission does take place for arbitrarily small values of the intensity. The effect of lowering the intensity is simply to decrease the photo-current without stopping the emission altogether.

Variation of stopping potential with frequency; no photoelectric emission takes place if the frequency is less than the threshold value v0, however large the intensity may be.

The Role of Photons in Photoelectric Emission

The classical theory could not account for all these observed features of photoelectric emission. For instance, classical theory tells us that whatever be the frequency, photoelectric emission should occur if the intensity of radiation is high enough since, for a high intensity of radiation, electrons within the emitting material should receive sufficient energy to come out, overcoming their binding force.

It was Einstein who first gave a complete account of the observed features of the photoelectric effect by invoking the idea of the photon as a quantum of energy, as introduced by Planck in connection with his derivation of the black body spectrum formula.

While the photons in the black body radiation were the energy quanta associated with standing wave modes, similar considerations apply to propagate radiation as well. Indeed, the components of electric and magnetic field intensities of propagating monochromatic electromagnetic radiation vary sinusoidally with time. Once again, a propagating mode of the field can be looked upon as a quantum mechanical harmonic oscillator of frequency, say, v. The minimum value by which the energy of the radiation can increase or decrease is once again hv, and this increase or decrease can once again be described as the appearance or disappearance of an energy quantum, or a photon, of frequency v. Such a photon associated with a progressive wave mode, moreover, carries a momentum just like any other particle such as an electron (by contrast, an energy quantum of black body radiation has no net rate). The terminologies for energy and momentum of a photon of frequency v are the de Broglie relations by now familiar to us: where λ stands for the wavelength of the propagating monochromatic radiation and where only the magnitude of the momentum has been considered.

When monochromatic radiation of frequency v is made to be incident on the surface of a metal or a semiconductor, photons of the same frequency interact with the material, and some of these exchange energies with the electrons in it. This can be interpreted as collisions between the photons and the electrons, where the power of the photon engaged in a crash is transferred to the electron. This energy transfer may be sufficient to knock the electron out of the material, which is how photoelectric emission takes place.

Bound Systems and Binding Energy

A metal or a semiconductor is a crystalline material where a large number of atoms are arranged in a regular periodic structure. Electrons in such material are bound with the entire crystalline structure. In this context, it is essential to grasp the concept of a bound system. For instance, a small piece of paper glued to a board makes up a set system, and it takes some energy to tear the piece of paper away from the board. If the power of the network made up of the paper separated from the board be taken as zero (in the process of energy accounting, anyone energy can be given a pre-assigned value, since power is undetermined to the extent of an additive constant), and if the energy required to tear the paper apart be E, then the principle of conservation of energy tells us that the power of the bound system with the paper glued on to the board must have been −E since the tearing energy E added to this initial energy gives the final power 0.

As another instance of a bound system, consider a hydrogen atom made up of an electron "glued" to a proton by the attractive Coulomb force between the two. Once again, it takes energy to knock the electron out of the atom, thereby yielding an unbound electron separated from the proton. The power of the divided system, with both the proton and the electron at rest, is taken to be zero by convention, in which case the expression gives the energy of the bound hydrogen atom with the electron in the nth stationary state. Notice that this energy is a negative quantity, which means that positive energy of equal magnitude is necessary to tear the electron away from the proton. This method of knocking an electron out of an atom is known as ionization. It can be accomplished with the help of a photon, which supplies the necessary energy to the electron, and the process is termed photo-ionization.

In an exactly similar manner, a hydrogen molecule is a bound system made up of two protons and two electrons. Looking at any one of these electrons, one can say that it is not bound to any one of the two protons but the pair of protons together. Indeed, the two electrons are shared by the pair of protons and form what is known as a covalent bond between the protons. Once again, it takes some energy to knock any one of these electrons out of the hydrogen molecule.

The minimum energy necessary to separate the components of a bound system is termed its binding energy. On receiving this amount of energy, the components get separated from each other without acquiring any kinetic energy in the separated configuration. If the bound system receives an amount of energy greater than the binding energy, then the extra amount goes to generate kinetic energy in the components. In this context, an interesting result relates to the situation when one of the components happens to be much lighter than the other. In this case, the extra energy is used up almost entirely as the kinetic energy of the lighter component.

Incidentally, when I speak of a bound system, I tacitly imply that it is to be looked at as a system made of two components. The same system may be looked at as one made up of more than two components as well. For instance, in the example of the piece of paper glued onto the board, the components I have in mind are the paper and the board. But, given a sufficient supply of energy, the board can also be broken up into two or more pieces, and then one would have to think of a system made up of more than two components. Indeed, the board and the piece of paper are made up of a large number of molecules, and the molecules can all be torn away from one another. Similarly, all of the two electrons and the two protons making up the hydrogen molecule can be pulled away from one another, for which a different amount of energy would be required as compared to the energy required to yield just one electron separated from an ion. This latter, we term the binding energy of the electron in the hydrogen molecule.

Chapter 13. Wave-Particle Duality

Despite the typical wave behavior observed in the experiment, light still has the properties of particles as well. Firstly, as we already know, it is divided into quanta called photons. Secondly, it can leave shadows and patterns resulting from holes in the wall. Additionally, if only one slot is left open during the experiment, one neat band is formed opposite this slit, resulting from the particle's flux. The question was, how do we understand this dual nature of light, and how can we describe it? Why does light act like waves in one case and like particles in another?

At first, scientists tried to explain where the waves come from using the water analogy. Light is a collection of particles, just like a water body is a collection of water molecules, and a set of particles, like a large number of water molecules in water bodies, can form waves. Therefore, each quantum, each photon (single unit of light), must be a particle. It was easy to test in a new version of the two-slit experiment, but this experiment did not confirm expectations! When the photons were fired one by one (for example, one per minute) towards the slit, each one appeared on the second wall, not in front of either slit, but randomly in one of those scattered places where interference bands had appeared in the standard version of the experiment! The wall retained a visual trace of the light particle (this was not just a wall, but a special screen that retained all the light traces), and over time, with each subsequent photon, an interference pattern was increasingly evident.

Why do the individual photons not appear on the screen directly in front of the slits? Why don't two stripes form on the screen?

This behavior of every single photon was completely unexpected and incomprehensible. The individual photons could not interact with any other particle because the photons were fired one at a time with an interval in between, much more spaced apart than light quanta usually travel. However, the final position of each photon at the screen was the result of interference. At the same time, upon getting to the screen, every single photon still left a point trace, just as would be expected of a particle.

These results cannot be explained in terms of the reality known to us. It seems evident that those photons could somehow move between the state of either particles or waves. No one had ever encountered anything like this before. The facts of quantum reality that we will discuss next are no less weird, but these are real facts of the microcosmic world since all of these are the results of observations and controlled experiments. Is it therefore so unusual that this field is difficult to understand since it consists of a range of completely new phenomena from the microcosmic level of reality that cannot be explained by the notions customary to us and even contradict them? Our perceptions of what is generally possible turned out not to be final since, until then, we had been dealing only with our macro world with its simpler laws and interconnections between the facts. The new facts related to the micro world, which have no analogs in the macro world, seem at least strange and often even unbelievable. In the language of science, this refers to the difference between quantum physics and regular classical (Newtonian) physics, known to a certain extent to each of us from school and everyday experience.

The fact that individual photons exhibit properties of both particles and waves proves that our strict division of reality into particles and waves is not entirely correct. Things aren't as simple as we thought. It turns out that particles and waves are concepts that may relate to the same phenomenon (for example, waves of radiation and photons of radiation. The term "photon" is used not only in visible light). But what about this solid matter, which consists of particles?

Solid matter consists of atoms. An atom consists of particles, electrons, protons, and neutrons (the latter two consist of even smaller particles, namely, quarks). Further experiments showed wave properties displayed by electrons, neutrons, and even entire atoms and molecules! Everything that makes up what seems to be solid matter acts like waves as well! Each particle of matter can "blur" its position. This dual nature of the whole of reality is called wave-particle duality. Everything is made of particle waves.

But why does everything sometimes act like particles and sometimes like waves? This is hard for people to imagine, even today. Therefore, it is accepted as a given, as part of a new set of facts about an earlier unknown reality, and scientists successfully work with it.

Chapter 14. The Theory of Relativity

Without exception, the peculiar relativity theory is one of the most influential developments during the history of science, and 2nd only to the observation by Newton of mechanics laws in their significance to physics. Despite this, peculiar relativity remains little known & there is a lot of confusion regarding the topic on the internet & in newspapers. Largely undeserved notoriety for being too hard for certain persons to grasp does not support this.

The simple concepts aren't that hard to grasp. By following a clear route through the evolution of physics from Galileo, this essay would clarify some of those fundamental theories, explaining why laws of physics since they were known in the 19th century would have to be modified, demonstrating how specific relativity emerged from that change, and discussing some of the implications of that latest theory.

As relativity theory emerged in the early 1900s, it upended science for decades and offered physicists a modern view of space and time. Isaac Newton saw space and time as defined, but they became dynamic and malleable in the new picture given by special relativity and general relativity.

Special relativity

Einstein said all observers would measure the light speed to be 186,000 miles per second, regardless of how fast and direction they move.

This theory inspired the comedian Stephen Wright to ask: "When you're in a spacecraft that's flying at the speed of light, you switching on the headlights, will something happen?" The reason is the headlights switching on naturally, but only from someone's viewpoint within the spaceship. To those waiting outside watching the ship pass past, the headlights do not seem to click on: light is coming out because it is moving at the same height as the starship.

These conflicting versions emerge since different observers do not have the same rulers and clocks — the items which define time and space. If, as Einstein had said, the speed of light is to be held steady, so time and space cannot be absolute; they must be relative.

For example, a 100-foot-long object moving at 99.99 percent light speed would look to a stationary spectator one foot long, but it will stay its usual length for those on board.

Maybe much weirder, the quicker one moves, the more time passes slowly. When a twin flies to a distant star in the racing spacecraft and then returns, she would be younger than her sibling who lived on Earth.

Weight also relies on velocity. The earlier an entity goes, the more he is huge. Of reality, no spacecraft can ever touch 100% of the speed of light because it would expand to infinity by its mass.

This relationship between mass and speed is also represented as a relation between mass and momentum: $E = mc^2$, where E is momentum, m is time, and c is the velocity of light.

What is general relativity?

Essentially that is the momentum principle. The fundamental principle is that instead of being an invisible force drawing objects to each other, gravity becomes a curving or spatial warping. The larger an entity, the more it distorts the surrounding area.

Einstein was not designed to disrupt our view of time and space. By incorporating acceleration, he moved on to generalize his theory and noticed that this skewed the form of time and space.

To keep to the illustration above: assume that the spaceship accelerates by shooting its propellers. All aboard should adhere to the ground as if on Earth. Einstein believed the power we call gravity was inseparable from being in a moving car. It was not so groundbreaking in itself, but as Einstein found out the complex math (it took him ten years), he realized that space and time are bent relative to a large body so that curvature is what we feel like the power of gravity.

In conclusion, special relativity is a series of equations that connect the way objects appear in one frame of reference to how they appear in another — the extension of time and space, and the change in mass. Nothing more complex than high school maths is included with the calculations.

General relativity becomes highly complex. The "field equations" explain the relationship between mass and space curvature and time dilation, which are usually taught in University Physics courses at the graduate level.

Tests of special and general relativity

A lot of experiments have verified the relevance of both special and general relativity in the last century. In the first significant general relativity study, in 1919, observers tested the deflection of light from distant stars as the starlight traveled through our sun, showing that gravity literally distorts or curves space.

In 1971, by putting specifically calibrated atomic clocks in airliners and sending them around the globe, physicists tested all aspects of Einstein's theory. A test of the timepieces after the planes landed found that the clocks on board the airliners were working a little bit slower than (less than one-millionth of a second) the clocks on the ground. The difference arose from the speed of the planes (a special effect of relativity) and their greater distance from the middle of the gravitational field of Earth (a general effect of relativity).

The observation of gravitational waves — tiny ripples through the space-time fabric — was yet another proof of general relativity through 2016.

Relativity in practice

While the concepts behind relativity sound abstract, the theory has profoundly affected the real world.

Of instance, nuclear power plants and nuclear bombs will be impossible without the understanding that matter can be turned into electricity. And our satellite network's GPS (global positioning system) needs to account for the subtle effects of both special and general relativity; if they did not, they would yield results that were off by several miles.

Understanding this universal phenomenon led Einstein to formulate the equivalence principle, according to which a gravitational field is locally equivalent to a field of acceleration. To obtain this principle, he drew upon a fundamental property of gravitational fields already brought to light by Galileo and included in Newton's equations: the acceleration communicated to a body by a gravitational field is independent of its mass.

After the development of special relativity, the need to generalize the theory seemed inevitable for multiple reasons. Relativist unification was far from complete. If the mechanics of free particles and electrodynamics finally satisfied the same laws, it was not the case for Newton's theory of universal gravitation, otherwise the top showpiece of classical physics. The equations of Newton are invariant under the classical transformation of Galileo, but not under those of Lorentz. Thus physics remained split in two, in contradiction with the principle of relativity, which necessitates the validity of the same fundamental laws in all situations.

Moreover, the Newtonian theory is based on certain presuppositions in contradiction with the principle of relativity: it is so with the concept of Newtonian force, which acts at a distance by propagating instantaneously at an infinite speed. The construction of a relativist theory of gravitation thus seemed to Einstein (and other physicists) a logical necessity.

Another problem was just as serious: the relativist approach explicitly gives itself the problem of changes in reference systems and their influence on the form of physical laws. But the answer provided by special relativity is only partial. It only considers frames of reference in uniform translation, at constant speeds concerning one another. However, the real world constantly shows us rotations and accelerations, from the fact of the multiple forces which are at work (such as gravity), or inversely, causing new forces (such as the forces of inertia).

What are the laws of transformation in the case of accelerated frames of reference? Why would such frames of reference not be as valid for writing the laws of physics as inertial frames of reference? The answer is that such a question requires a generalization of special relativity.

The originality of Einstein's approach had been, in particular, to bring together two problems, that of constructing a relativist theory of gravitation and that of generalizing relativity to non-inertial systems, into a single endeavor. The equivalence principle made this unity of approach possible: if the field of acceleration and gravitational field are locally indistinguishable, the two problems of describing changes in the coordinate systems, including those which are accelerated and those which are subject to a gravitational field, boil down to a single problem. But such an approach is not reducible to "making relativist" Newtonian gravitation. While certain physicists could hope, at the time, that the problem of Newton's theory could be solved by a simple reformulation, by introducing a force that propagated at the speed of light, it is the entire framework of classical physics that Einstein proposed to reconstruct with general relativity. Better yet, it was a new type of theory which he developed for the first time: a theory of a framework (curved spacetime, now a dynamic variable) in connection with its contents, and no longer only a theory of "objects" in a rigid preexisting framework (as was Newton's absolute space).

Why such a radical choice? Doubtless, because special relativity itself was unsatisfactory on at least one essential point: the spacetime which characterizes it, even if it includes in its description space and time which is no longer absolute taken individually, remains absolute when taken as a four-dimensional "object." However, inspired in particular by the ideas of Ernst Mach, Einstein had come to think that an absolute spacetime could have no physical meaning, but rather, that its geometry should be in correspondence with its material and energetic contents. Thus a reflection on the problem of inertial forces, which had caused Newton to introduce absolute space, led Einstein to the opposite conclusion.

The Problem Of Inertial Forces

The existence of inertial forces acutely poses the problem of the absolute or relative nature of motion and, ultimately, of spacetime. The ideas of Mach in this area had a deep influence on Einstein. For Mach, the relativity of motion did not apply solely to uniform motion in translation; rather, all motion of whatever sort was by essence relative (Poincaré and, long before him, Huygens had arrived at the same conclusions).

This proposition can seem in contradiction with the facts. If it is clear, since Galileo, that it is impossible to characterize the state of the inertial motion of a body in an absolute manner (only the speed of a body concerning another has physical meaning), it seems different in the case of accelerated motions. Thus, when one considers a body turning about itself, the existence of its rotational motion seems to be able to be felt in a manner intrinsic to the body. No other body of reference is needed: it is enough to verify whether or not a centrifugal force appears, which tends to deform the rotating body.

In reconsidering the thought experiment of Galileo's ship, the difference between inertial movement and rotational motion becomes heightened. No experiment conducted in the cabin of a ship traveling in uniform and rectilinear motion concerning the Earth is capable of determining the existence of the boat's movement: as Galileo understood, "motion is like nothing." Relative motion can only be determined by opening a porthole in the cabin and watching the shore pass by. But now, if the boat accelerates or turns about itself, all the objects present in the cabin will be pushed toward the walls. The experimenter will know that there is movement without having to look outside. Thus, accelerated motion seems definable by a purely local experiment.

It is such an argument that caused Newton to allow that one can define an absolute space, in opposition with Leibniz (then Mach), for whom defining a space independently of the objects it contains could not have meaning.

Mach proposed a solution to the problem completely different from Newton's. Starting from the principle of relativity of all motion, he arrived at the natural conclusion that the turning body, within which there appear inertial forces, must turn not concerning a certain absolute space, but concerning other material bodies. Which ones? It cannot be tight bodies, of which the fluctuations of distribution would provoke observable fluctuations of inertial systems. This is unacceptable since it is easy to verify the coherence of these systems over great distances. Thus, if we look, motionless concerning the Earth, at the night sky, we do not see the stars turning.

Nevertheless, if we turn about ourselves, we feel our arms spreading out due to inertial forces, and, in raising our eyes toward the sky, we can see it turn. This was the initial observation of Mach: it is within the same frame of reference that the arms are raised, and the sky turns, and this will be true for two points of the Earth separated by thousands of kilometers. Mach suggested, then, that the common frame of reference is determined by the entirety of the foreign matter of bodies "at infinity," of which the

78

cumulative gravitational influence would be at the origin of inertial forces. In other words, the body would turn concerning a frame of reference, not absolute, but universal. An absolute motion would be defined, independently of all objects. However, Mach argued, all motion is relative, remaining defined for an "object," even if this object is the universe in its entirety.

Relativity Of Gravity

If an observer descends in free fall within a gravitational field, they no longer feel their weight, which means that they no longer feel the existence of this field itself. This remark, which can now seem obvious to us—we have all seen, on television or in movies, astronauts in weightlessness floating in their ship, and the objects that they drop going away from them at a constant speed—is nevertheless revolutionary, for it implies that gravity does not exist, that its very existence depends on the choice of a frame of reference.

He thus distanced himself from the former concept of gravity. What can be more absolute than a gravitational field in the Newtonian model? Gravity had been recognized by Newton as universal; here indeed was a physical phenomenon of which the existence does not seem to be able to depend on such and such a condition of observation.

However, if we allow an enclosed area to fall freely within a gravitational field, and then put in motion a body at a certain velocity to this area, the body will move in a straight line at a constant speed for the walls of the enclosure; a body initially immobile (again, to the walls) will stay thus during the movement of the enclosure's fall. In other words, all experiments that we can perform there would confirm that we are in an inertial frame of reference! Thus, gravity, however universal it is, can be canceled out solely by a judicious choice of the coordinate system. What Einstein understood in 1907 was that even the existence of gravity was relative to the choice of the coordinate system.

Relativity and Quantum Mechanics

In its ordinary form, quantum theory abandons some of the basic assumptions of the theory of relativity, such as the principle of locality used in the relativistic description of causality. Einstein himself had considered the violation of the principle of locality to which quantum mechanics seemed to be absurd. Einstein's position was to postulate that quantum mechanics, although consistent, was incomplete. To justify his argument and his rejection of the lack of locality and the lack of determinism, Einstein and several of his collaborators postulated the so-called Einstein-Podolsky-Rosen (EPR) paradox, which demonstrates that measuring the state of a particle can instantaneously change the status of your linked partner, although the two particles can be at an arbitrarily large distance. Modernly, the paradoxical result of the EPR paradox is known to be a perfectly consistent consequence of the so-called quantum entanglement.

Chapter 15. Quantum Field Theory

The quantum theory development was not finished with quantum mechanics. Because it has its limits, it doesn't contain a particular view of relativity. Therefore, it only applies to objects that move much slower than the speed of light. What is valid for the electrons in the atoms and molecules, thus quantum mechanics fits wonderfully. The photons, which always move at the speed of light, are not covered by the Schrödinger equation. As already mentioned, the English physicist Paul Dirac has developed the abstract version of quantum mechanics. In 1928 he succeeded in integrating the particular theory of relativity into the Schrödinger equation. This is the Dirac equation, which I have given as an impressive example (keyword antiparticles) that we discover mathematics and thus the laws of physics and not invent them. But even with the Dirac equation, the photons could not be described. Moreover, like the Schrödinger equation, it only applies to a constant number of particles. The particular theory of relativity, however, makes its creation and destruction possible. Both the photons and the variable particle number required an entirely new concept, that's quantum field theory (QFT). However, the passage of quantum mechanics from quantum field theory was only a small step compared to the enormous revolution from classical physics to quantum mechanics.

Their development began already in the '20s, parallel to quantum mechanics. Paul Dirac, Werner Heisenberg, and Pascual Jordan were significantly involved. You already know them. The Italian physicist Enrico Fermi (1901—1954) and the Austrian physicist Wolfgang Pauli (1900—1958) made contributions. But the development quickly came to a standstill because there were absurdities in the form of infinitely large intermediate values that could not be eliminated for a long time. It was not until 1946 that people learned to deal with them. The first quantum field theory emerged around 1950; quantum electrodynamics (QED) was the Maxwell equation's quantum version. The American physicist Richard Feynman (1918—1988), a charismatic personality, played a decisive role in its development. He tended to bizarre actions, for example, regularly drummed in a nightclub. And he was involved in solving the challenger disaster in 1986. Richard Feynman is the only physicist who still gave lectures for beginners when he was already famous. That was in the early '60s. They have also been published in book form and are still widely used today. The German textbooks on physics are, as expected, factual and dry. The Feynman textbooks are more "relaxed "and contain much more text. My impression; for the first contact with physics, they are less suitable, but they improve its understanding if one has already learned some physics. Quantum electrodynamics is the foundation of everything that surrounds us. Real chemistry and thus also biology follow from it. But even with it, the development of the quantum theory was not yet complete. Because two more forces were discovered, the strong and the weak point. They only play a role in the atomic nuclei, so we don't notice them. We only see the gravitational and the electromagnetic force. Even though large bodies are always electrically neutral, quantum electrodynamics also plays an essential role in everyday life. The matter is almost empty. Accordingly, if two vehicles collide, they should penetrate each other if it wasn't for QED. The strong and weak forces led to the development of two other quantum physics theories. Whereby the QFT of the weak point could be combined with the QED. For the sake of completeness, the QFT of the vital force is called quantum chromodynamics (QCD). The QFTs of the three points that can be described with them are combined to form the standard particle physics model. It represents the basis of modern physics.

In the QFTs, each type of elementary particle is described as a field in which particles = quanta are created and destroyed. This is the core of the QFTs. What are the elementary particles? Matter consists of molecules that consist of atoms that can be separated into a nucleus and electrons. The electrons are called elementary particles because they cannot be further divided according to today's knowledge. The core comprises protons and neutrons, each consisting of three quarks, or more precisely, of different mixtures of up and down quarks. They are elementary particles. The ordinary matter thus consists of three other elementary particles. But there are many more matter elementary particles, 24 in total, and that's not all either, because three of the four different forces also consist of elementary particles. Only the gravitational force cannot be integrated into the concept of the QFTs.

And there is another elementary particle, the Higgs particle, discovered at the Large Hadron Collider (LHC) in Geneva in 2012. It plays a unique role; I won't go into it further.

Since the general theory of relativity cannot be integrated into the standard particle physics model, there must be a more fundamental theory. String theory was a promising candidate, but the hope is fading. Will the more fundamental theory, if it is ever found, be the "theory of everything" (TOE)? So, will it herald the end of theoretical physics? Some believe this, such as the recently deceased physicist Stephen Hawking (1942—2018). So far, however, every physical theory always has something inexplicable in it. One can hope that we will find an approach that explains everything. But I doubt that we will ever succeed. Therefore, even in the far future, theoretical physicists are looking for new theories.

Quantum field theory (QFT) integrates special relativity and quantum mechanics, thus providing a description of subatomic particles and their more accurate interactions than quantum mechanics.

In the quantum field, theory particles are viewed as excited states of the underlying physical area; excitations of the Dirac field manifest themselves as electrons; excitations of the electromagnetic field give rise to photons. Moreover, particle interactions, such as those occurring in high-speed collision experiments in an accelerator, are described by perturbative or iterative terms among the corresponding quantum fields. At each level of iterative approximation, these interactions can be distilled into a diagram, known as a Feynman diagram; the chart shows that particle interactions occur via an exchange of mediator or force carrier particles. Quantum electrodynamics (QED) is the first quantum field theory to be developed; quantum electrodynamics combines the relativistic theory of electrons (encapsulated in the Dirac equation) quantized electromagnetic field in the form of a Lagrangian.

The quantum electrodynamics Lagrangian with Feynman calculus aid leads to scattering amplitude calculations; these calculations represent the most accurate physical predictions in the entire physics.

Suffice it to say that we would not understand the particle physics we have today without the quantum field theory. It is natural to take up the quantum field theory once one has gone through quantum mechanics and relativity theory.

Simultaneously, QFT is considered one of the most challenging subjects in modern physics; the theory seems contrived and unnecessarily complicated at times, mixing a disparate set of ideas in a series of detailed calculations.

Chapter 16. Particle Physics

If chemistry has its Periodic Table of Elements, quantum physics has its Standard Model of Elementary Particles—a set of formulae and measurements that describe elementary particles and the way they interact. Where the Periodic Table of Elements categorizes atoms based on their specific characteristics, the Standard Model categorizes elementary particles in two main groups; Fermions and bosons.

Developed at the beginning of the 1970s, the Standard Model came as a means by which quantum physicists attempted to standardize what was already known about particles and forces. This effort had the goal of furthering research in the field and helping other scientists refer to a standard measurement of the elementary particles rather than to a disparate cumulus of theories and theses.

In addition to presenting what was already known to the scientific community (and, as such, what was already accepted by it), the Standard model also helped predict the existence of particles that were not discovered yet (such as the (in) famous Higgs boson).

The Standard Model is considered to be one of the best-formulated theories in particle physics. However, it still has gaps that scientists are trying to uncover—like how general relativity's approach on gravity can be integrated into quantum theory or to explain why there is more matter than antimatter in the universe.

Fermions

According to the Standard Model, fundamental particles are categorized into groups that are related to each other according to certain rules. For instance, two fermions cannot be in the same place at once. This rule allows them to build, block upon block, everything (from atoms to the shoes you're wearing and the planets of the universe).

There are two categories of fermions:

- **Quarks:** Which can combine into protons and neutrons. For instance, a proton consists of two up quarks and one down quark, which are connected by a strong nuclear force.

- **Leptons:** Which include electrons, muons, taus, as well as neutrino electrons, neutrino muons, and neutrino taus. These neutrino particles are very interesting because they are barely noticeable, but they might still have a heavy influence on the world.

Bosons

If fermions make up the world, bosons act as liaisons between different fermions. They are, in some ways, mediators between the forces that act on the matter by either binding or repelling it.

Bosons explain why we can't do certain things—such as why the light comes in the colors of the rainbow, why human beings can't walk through walls, and so on.

Breaking the mystery of bosons would be a tremendous achievement for mankind. We might not be daring enough to think of X-Men-like superpowers that will allow us to walk through walls (to circle back to the aforementioned example), but managing to control certain bosons would give us a power that would have been considered superhuman just a couple of hundreds of years ago.

In the bigger family of bosons are included:

- Photons (which communicate the electromagnetic force)
- Gluons (which deal with the strong nuclear force needed to bind quarks together and create protons and neutrons)
- W & Z bosons (which have to do with the weak nuclear force)

- Higgs boson (which is meant to explain why some particles have mass when certain conditions are met).

Of course, this is all a very short version of the entire story. As you can imagine, the information in the Standard Model of Elements lays at the foundation of future science.

For instance, do you remember how years ago everyone went hysterical about a scientific program in Switzerland that was risking creating a "black hole" that would absorb us all? What they were trying to do was to isolate the Higgs boson, which was a beginning in the understanding of an array of concepts, including (but not limited to) dark matter.

Fascinating, right? And this is just a very, very small piece of the entire ensemble of quantum physics.

Chapter 17. String Theory

Throughout this book, you've heard of the term "string theory," but you might not yet have a precise idea of what it means. As it turns out, this is the case for many people, and not by their fault, either—string theory is just really, really confusing.

String theory is so named because it was created to describe the nature of quantum objects as something like a vibrating string, and those objects were measured based on those vibrations. Without getting too deeply into it—string theory involves a lot of work with subatomic particles, which are the smaller particles that make subatomic particles, like quarks, string theory was essentially formulated to help explain strange interactions that sometimes happened between some of the known subatomic particles. These particles would sometimes act like they were bound together by strings. Thus, string theory was formed. Now, one of the most interesting parts of this early string theory was that these weird, string-based interactions worked out some mysterious math. For one, the vibrations in the strings predicted the existence of a certain particle; the graviton. It's the only quantum theory that has been able to do this so far successfully. Theoretically, a graviton is a particle that causes gravity, and one of string theory's biggest advantages here is that its graviton is shaped like a donut. That may sound strange, but that same shape prevents many of the mathematical anomalies that arise when trying to envision gravity as a particle. However, with today's science, we still can't prove that the graviton even exists. Gravity still presents many mysteries to us.

The string theory was able to describe gravity as its quantum particle this way successfully. However, string theory has many limitations that keep it from being an attractive (or logical) solution to the quantum universe's mysteries. For one, the final proposal of string theory requires ten dimensions to function properly, but we've so far only observed four dimensions in our existence. To remedy that, physicists and scientists alike have tried to make string theory work in its prerequisite ten dimensions, then remove the extra dimensions that don't apply to our universe. However, none have yet been successful. Besides, there are several other versions of string theory—there isn't just one. Bosonic string theory was one example, and it required twenty-six dimensions!

Ed Witten managed to combine many of the varied string theories into M-theory. However, M-theory required eleven dimensions, so while it managed to unify many of the past string theories into one neat package, it still didn't seem anywhere near reasonable science.

In addition, the "theory of everything," as its name implies, is an all-encompassing theory. It explains in a way or another how quantum physics and quantum particles worked. Although it is one of the main goals of today's physicists and scientists, we have not yet formulated a fully working theory. However, string theory makes a compelling case for becoming or at least supplementing a potential theory of everything.

It seems that string theory makes a very neat argument to explain the nature of particles. Imagine a guitar string in slow motion; the string moves in a wave shape, just like light or sound. The tension (how taut the string is pulled) of the string also controls what note any string on a guitar plays. If you pluck the string, it will vibrate. This is the prevailing theory behind the strings; each string has a different frequency (like the note of a guitar string) that it vibrates at, and each string also has a different length, which changes the number of notes that the string could play. Think about how, when playing the guitar, moving your fingers down the neck, shortening the string, will play higher notes while moving your hand up the neck will lengthen the string and play lower notes. Interestingly enough, the intervals between these possible "notes" are also defined by a familiar variable; Planck's constant. Unfortunately, as convenient as this might seem, there's one question that scientists seem to be unwilling (or unable) to answer about string theory. What are the strings made of? Answers that have been tried are "pure mass" (whatever that means), irregularities in the fabric of reality, and the "energy" of "existence." The best answer, however, is, "The answer is irrelevant because they're there."

One of the biggest downfalls of string theory is that, because it's so complicated and exists in so many additional dimensions (that we can't even find), we can't test it. However, we also can't rule it out as entirely wrong, either. This creates a bit of a catch twenty-two; should scientists and physicists keep pouring man-hours and research into a theory that we technically can't even test, or should we just give up on it and risk sidelining a potential theory of the universe? There's no good answer to this question, but it's one of the pitfalls that plague string theory and those interested in it.

Despite its shortcomings, though, many scientists still believe that string theory is the way things are—or, at least, it's a step to figuring it out. However, until we find six or seven more dimensions, we'll have to wait on that.

Chapter 18. The Law of Conservation of Energy

What Is the Law of Conservation of Energy

The law of conservation of energy is a fundamental principle in physics. It states that the total amount of energy in the universe is constant, and this energy can be divided into three components; matter, kinetic, and potential. Energy cannot be created or destroyed, only transformed from one form to another.

It is a common misconception that energy can be "lost" during a transformation. Energy must not be thought of as a commodity that is stored in a given system but rather as a process by which certain effects are achieved. In the case of Lossless energy conversion, when, for example, thermal energy is converted to kinetic energy when an object falls from a height, the thermal energy has been converted into kinetic energy, and thus, no loss in total system potential energy has occurred; this phenomenon can be observed when lifting an object at constant velocity.

Some examples of systems where the law of conservation applies:

- A photon (or another particle) is absorbed by an atom, and the energy is temporarily given to the nucleus.
- An electron in orbit around a nuclear atom has its energy completely absorbed by the nucleus.

There are four possible outcomes when an energy transformation takes place:

- **A new form of energy (or forms) is given to the system:** The original form of the system's energy has been conserved, and this result has been achieved using input heat energy.

- **The original form of the system's energy has been conserved:** The energy is not changed.
- **The energy is transformed from one form to another:** A smaller amount of input heat energy is needed to get the same amount of output heat (or work).
- **The system loses heat (or work) in the transformation:** A larger amount of input heat (or work) is needed to get the same amount of output heat (or work).

A system is said to be in equilibrium if its energy has been transformed into the form of another form of energy and the total amount of energy in the system remains constant. For example, a light bulb in a room keeps its temperature constant and does not lose or gain heat in any way, so the light bulb is at equilibrium with its container.

The law of conservation is one of four laws that physicists use to simplify their thinking about the world around us. By using only three laws, they can solve all kinds of complex problems that would otherwise require large computational resources, such as supercomputers. (The other three laws are conservation of momentum, conservation of angular momentum, and conservation of electric charge.)

The law of conservation is often stated in mathematical terms in the form of the equation E=mc2. This equation shows that energy can be transformed into different matters at a fixed rate; this means matter can be created from energy. There are many claimed variations on this theme, which are generally useless. Most examples attempt to state that material may be created from energy by the action of some unknown force or law or that matter can disappear into or out of existence.

Application to Quantum Physics

Quantum physics explains that atoms and nuclei are constantly undergoing radioactive decay to other types of atoms and nuclei.

This process may result in the release of large amounts of energy and is one way in which the Sun and the stars produce energy. Scientists use Einstein's equation $E=mc2$ to express this process.

This equation also explains the connection between matter and energy. Mass is simply a sum of energy. The conversion between mass and energy corresponds to the binding energy of the nucleus in an atom or a nucleus or to the difference in binding energies when different elements are formed by nuclear fusion.

The Law of Conservation of Energy has two apparent limitations:

- "There is no evidence that mechanical, thermal, chemical, nuclear, or electrical energy can be created out of nothing."
- "It is impossible to extract electric energy completely from any single mechanical, thermal, chemical, nuclear, or other sources alone."

This is a serious limitation because energy is essential to the functioning of the Earth's core and for humans to survive. It is obvious, therefore, that if this limitation was true, no life could exist on Earth.

This limitation is not true at all. The law of conservation of matter applies to solid material objects, not extended systems such as living organisms. In a low-energy system such as an atom or a nucleus, the total mass and energy content of which is conserved in any process, the total mass and energy contents remain constant during that process. Thus, it can be said that matter and energy are completely inseparable, and one cannot be created without the other.

Chapter 19. Thermodynamics and Kinetic Theory

What Are Thermodynamics

Thermodynamics is the study of heat and its relation to energy, work, entropy, temperature, and pressure. Much like in Newtonian mechanics that only focuses on the motion of bodies at rest or in uniform motion due to unopposed forces (equations for velocity are independent of forces), thermodynamics only relies on energy.

What Is Kinetic Theory

Kinetic theory is a branch of physics that studies the behavior of gases. It explains how kinetic energy is related to temperature and pressure for a given volume of particles, as well as how different particle arrangements result in different pressures.

What Is the Difference Between These Two Theories

The difference between them is that thermodynamics studies changes, while kinetic theory explains the properties of systems.

How does this relate to our discussion on the relation between temperature and entropy?

Entropy is a reservoir of randomness that grows spontaneously. By studying randomness, thermodynamics shows how energy is converted into heat. As we will see later in our discussion on entropy and temperature, there are many ways in which heat can be converted into work by various devices. Still, all of them involve some level of randomness.

Thermodynamics and Kinetic Theory in Quantum Physics

The two theories above are mostly classical. They focus on macroscopic events. However, quantum physics is based on the idea of discrete particles.

The fact that the number of particles cannot be defined in terms of continuous variables (like temperature or pressure can be), but only in terms of discrete quantities like energy, means that temperature and pressure cannot be meaningful concepts in quantum physics.

This means that while thermodynamics is important to know, kinetic theory cannot be used to make predictions about the behavior of gases at all because there are no gas particles in quantum physics for it to happen with. Only probability has meaning for strange wave-particle duality systems like that.

There are some ways to use kinetic theory to study gas particles, such as the kinetic theory of gases, which has been used for a long time to study the flow of fluids. However, this was not a formal theory.

Instead, statistical mechanics use a combination of first principles and kinetic theory. It is a rather complicated topic, but it can be simplified as follows.

In statistical mechanics, the state of a system is completely defined by its volume and energy distribution within that volume. The two quantities are called entropies. In thermal equilibrium systems, entropy stays constant as heat flows from hot things to cold things or cold things into hot things. The entropy of a system is also called its thermodynamic entropy.

It has been argued that thermodynamics does not explain the behavior of non-equilibrium processes because it does not explain how entropy diverges. Entropy is the degree to which an isolated system will tend to increase or decrease. If you think about it, this would mean that thermodynamics cannot explain why there are laws of motion in nature.

We can tell by looking at classical thermodynamics that there are laws of motion everywhere. They are simply not stated in terms of energy.

Chapter 20. Quantum Entanglement

One of the basic and often seen as a strange occurrence in quantum mechanics is the concept of "entanglement." Entanglement is a phenomenon where two quantum particles interact appropriately and how their states will depend on each other, irrespective of how far apart they are.

Einstein's empirical study on momentum and position measurements led him to this concept. Indeed, entangled particles can appear remote and distant, yet they have the same physical structure, which allows an interrelation and an interconnection.

Another model to consider genuinely in our modern days is the interconnection and the polarization of the entrapped photons. It's clear that when the entanglement of electrons occurs although particles are not connected, they are dependent upon similar movements and happenings. Concretely, this model isn't completely perceived by physicists, even though it is considered a fundamental standard of quantum physics.

The spin may have a positive or negative (upward or downward) value (or direction, known as its "sign"), and when the electrons are entangled, it means that the measurement will show that their spin signs are opposite. If the spin of one of the entangled electrons is determined, it is immediately known that the spin of the second electron has the opposite sign. In reality, this entanglement occurs, for example, when the particles are formed in a single process (in the case of experiments with photons, identical, entangled photons are produced in the process of decomposition of one photon twice their size) or when they are components of one system, such as electrons being components of one atom.

In the case of entanglement, can these characteristics be coordinated with the speed of light rather than instantaneously? There are no methods to measure time and speed with absolute precision. However, the instrument's precision is constantly improving, and modern experiments have shown so far that the speed of interaction between particles during entanglement exceeds the speed of light by at least 100,000 times! Scientists assume that if the speed of interaction during entanglement exceeds the speed of light (that much), then this interaction has infinite velocity, i.e., both particles acquire the exact characteristics simultaneously regardless of distance (non-locality).

For instance, if you measure two particles where they are both located in different countries and then measured at the same time, the measurement carried out in one country or location will absolutely and undeniably decide what the outcome in the other location will turn out to be. There is no theory to describe the correlation between these particles wherein they have certain states. It is known that the two states will remain indeterminate until one of the states is weighed; it is at this time that the states of each of the particles are then determined, not minding their distances apart. A lot of experiments have been carried out in the last thirty years using atoms and light to confirm this theory. The experiment carried out now still confirms the quantum prediction.

It is worthy to note that this does not in any way serve as a means to send signals that are faster than light since the measurement in one location says Oxford will determine the state of the particle in another location, maybe Harvard. This shows that the outcome of each measurement is random. The particle at Harvard cannot be adjusted to match with the result obtained for the particle at Princeton.

The connection between these two measurements can only be evident when the two sets of data are measured and compared, and this process has to be done at a speed that is quite slower than the speed of light.

Once a thing isn't forbidden, it is mandatory.

If a quantum particle is moving from point A to point B, it will take absolutely all of the path from point A to point B simultaneously. This usually includes paths that have unlikely events like electron-positron pairs that appear from nowhere and disappear suddenly. A field of Quantum physics referred to as quantum electrodynamics (QED) implies that every possible process should be studied, including the very unlikely ones.

The QED is not just a random process of guessing without any real application.

The prediction of the interaction between the electron and the magnetic field by QED correctly describes the interaction at 14 decimal places. As strange as this theory may seem, it remains one of the best-tested hypotheses in the history of science.

Chapter 21. Dark Matter

Many scientific theories and terms have been explained relating to Quantum Physics. As a branch of the study of physics, we understand that the knowledge gained is limited by our own ability to observe, measure, and explain it. Our observations, however, can create more questions and additional areas of study.

Explanation theories, such as the uncertainty principle and the paradoxes found within the earliest work and theories of Quantum Physics, provide many examples of how the field has grown and changed throughout the years.

Yet, in one area, it is clear that Quantum Physics measures something that cannot be seen, but the effects of this event require clarification. So, what is this area of study? It is dark matter.

To understand what dark matter is, the first thing we must do is explain the effects it has within the universe and how it was found at all.

Galaxies themselves move at an amazing rate of speed. The gravitation pulling caused by their matter that can be observed by scientists does not correlate to the amount of gravity that would be needed to hold these galaxies together. Simply put, they would have flown apart ions ago due to the forces created by the speed of their rotation. This was true even when the galaxies were found in their clusters. So how did these galaxies maintain their rotations?

Scientists pondered the data and attempted to figure out how this was possible. Eventually, the data itself began to reveal where the additional gravity might be coming from. Thus, scientists discovered dark matter, although it remains undetected at this time. So, what is dark matter?

This matter does not interact with electromagnetic forces as normal matter does. Imagine something that does not have any interactions with light. It doesn't absorb, reflect or put out any light, making it hard to find. Yet, the effects of dark matter can be observed by how it appears to work on visible matter, creating additional gravitational pull. All the matter we can observe and is part of our knowledge set regarding the universe, including what makes up the planets and stars, only accounts for approximately 5% of the universe. Researchers believe that dark matter appears to make up over 27% of the universe. How is this possible? Research has concluded that dark matter must weigh six times the weight of visible matter.

Scientists are not clear about what makes dark matter. Yet, just by discerning its existence, researchers now have a point of study. Theories have begun to pop up, including the idea that dark matter is made up of supersymmetric particles. These particles are hypothesized particles that are partners to those particles that are known through the Standard Model.

Understanding the Standard Model and Particles

Before we move forward, let's take a moment to explain the Standard Model. This is one of the fundamental theories that is used to explain how matter interacts with itself through four different forces. The simple building blocks of matter, referred to as fundamental particles, are directed by four central forces. When we take into account that Quantum Physics is explained by mathematical equations, it becomes apparent that the Standard Model has a few weaknesses.

One of the first weaknesses, so to speak, is that Standard Model only looks at particles and three of these forces. This model was established in the 1970s, and it has been able to explain a variety of results from various experiments, as well as predict the behavior of particles and another phenomenon very precisely, so scientists and researchers rely heavily on it. Additionally, because it works with particles, researchers can use it in their various studies of matter, especially elementary particles.

There are several types of elementary particles, the construction slabs of all matter. There are two simple categories of these particles, called leptons and quarks. Within each of these groups, because with Quantum Physics, everything is a part of something else, are six particles that relate to each other in pairs. These pairs are often referred to as generations. The most constant particles are also the lightest, resulting in the primary generation. However, the heftier and less constant of the particles can determine whether the particles end up as the second or third generation.

Stable matter within the universe is created of the particles that are classified as part of the first or primary generation because any of the particles from the other generations are too heavy, and they quickly decay into a more stable level. Quarks paired in these generations have additional names, such as up quark and down quark for the first generation, charm and strange quark for the second generation. Finally, the top quark and the beauty (bottom) quark make up the third generation.

Another distinguishing characteristic is that quarks come in three diverse colors (for lack of a better description), but they only blend in ways that end up forming colorless objects. For the leptons, the electron and its neutrino are from the first generation. The second is the muon and its neutrino. Finally, the tau and its neutrino create the third generation. For each of these pairings, the electron, muon, and tau have a substantial mass and an electrical charge. On the other hand, the neutrinos of these pairing are usually found to have little mass, and they are electrically neutral.

Within the universe, four essential forces have been identified, the weak, the strong, the gravitational, and electromagnetic forces. These forces not only have different strengths but the forces labor over a variety of ranges. The electromagnetic force works within an infinite range and is more intense than gravity. Strong and weak forces are found to dominate at the subatomic particle level and can only be found to be effective over a relatively short range.

Gravity is a weak force, but its range is infinite, similar to the electromagnetic force.

The weak force, in keeping with its name, is the frailest of all the forces, but due to its short range, it is also stronger than gravity. The strongest of the four classic interactions is a strong force.

Three of these forces are the result of force-carrier particles exchanges, belonging to the larger group referred to as bosons. Specific quantities of energy are transferred by particles trading bosons with other particles. Each of these forces has a corresponding boson. For example, the electromagnetic force has the photon. The weak force is associated with W and Z bosons, whereas the strong force has the gluon. Finally, gravity is associated with graviton. However, the graviton itself has not yet been found by researchers and scientists. Its existence is surmised based on the existence of the other force's bosons.

Within a Standard Model, the electromagnetic, weak, and strong forces are accounted for, along with the associated carrier particles. Scientists can explain how these forces work on various matter particles using this model. Yet, the force that we are most aware of and deal with daily, gravity, has not been included within this model. Why is this difficult to do? In part, it leads us back to Quantum Physics. As we have seen throughout our exploration, the various theories used to define a microdomain, along with the relativity theory that is used to explain the macro world, can be problematic to fit into one framework that explains them both.

Thus, they have not been able to be made mathematically compatible, especially in the Standard Model context. Yet within particle physics, their minuscule scale results in gravity having a weak or negligible effect at best. Bulky matter, on the other hand, is where gravity dominates. Some examples are the planets or bodies, animals or humans. Since a Standard Model is focused on the individual particle world, it works well in physics, although that it cannot include gravity as part of its modeling.

However, this model doesn't explain the complete picture of a subatomic realm. However, this particular model does give one of the best descriptions available so far. Yet it doesn't answer such important questions like what happened to the antimatter after such events as the big bang? Additionally, why are quarks or leptons created with three altered generations and their various scaled masses?

As these questions appeared, new experiments have been developed, and researchers are working hard to find the information that will fill in the gap in our knowledge of these forces and the physics of this subatomic world, using a hydron collider (LHC). Now let's take this information that we have gleaned on the particles in the Standard Model and take this information back to our discussion of dark matter.

Relating Particles to Dark Matter

At the LHC, experiments are being conducted to see if dark matter, which researchers infer is made of particles light enough to be made at the LHC, can be measured. While the light particles would escape undetected, they would drag away momentum and energy as the collision occurred. This is what the scientists could then measure to determine the dark matter existence.

In other theories, variations of dark matter appear, such as the supersymmetry and the theory of extra dimensions. Another theory suggests that dark matter exists in a parallel world of dark matter that behaves entirely differently from how we know matter to act. All these theories are potential areas of knowledge, and if any are true, then they can help expand the knowledge base of how galaxies hold themselves together.

Yet, in all this discussion, there is still almost 70% of the universe that is uncounted for. Researchers believe that dark energy is what makes up this portion of the universe in the form of vacuums and is distributed evenly throughout the universe. That distribution means that its gravitational effect is not felt on a localized level, but it is felt on a universal level. Dark energy creates a repulsive effect, thus being the focus of the expansion of the universe. Thanks to various measurements of this expansion, researchers have been able to confirm the existence of dark energy and its overall distribution. As you can see, dark matter provides a localized effect that helps the galaxies hold themselves together in addition to the gravitational pull of their mass. Thus, galaxies can continue to move at the rate of increased speed without spinning apart.

We have learned about a variety of pieces of the universe from a quantum perspective. As you can see, the various parts encompass both how the universe works and goes down to the building blocks of how it was made. While there are many areas of study that are still built only on working theories, researchers and scientists continue to build the base of their knowledge, thus proving and disproving various aspects of their theories and their equations.

Chapter 22. Black Holes

Einstein's theory is technically irreconcilable with quantum physics for several reasons that we have mentioned in this book. According to Einstein, black holes are regions of ultra-intense gravity within powerful space-time. They pull even light inside.

A black hole was first deliberated by a clergyman named John Mitchell in 1784. His ideas were dismissed mostly because, shortly afterward, the light was "discovered" to be a wave instead of a particle (at the time). Therefore, scientists were unsure if gravity would act on light "waves" instead of "particles" and mostly forgot about the concept. That is until Einstein brought them back in 1915 with his general relativity theory. From there, precisely how black holes are formed, how they work, and how they influence space and time have been hotly debated by many physicists. Scientists believe that time is so distorted in a black hole that, to an outside observer, time would appear to stop inside it. However, if a person were to fall into a black hole, time would proceed for them normally (resulting in a rather gruesome death).

Black holes are created in a few different ways. One way would be by violating Planck's law. It has to do with the Planck constant. Planck's constant also defines several measurements within quantum physics, and these measurements denote certain limits of the size that can achieve within the quantum universe. For example, if you tried to measure a particle's position with a laser at a greater accuracy than one Planck length (1.6×10^{-35} meters, which is very tiny), the laser's power would end up creating a little black hole. Ironically, the black hole made would be precisely the size of one Planck length. Since time and space are intertwined, a black hole can also be created when measuring a length of time less than one unit of Planck time (10^{-43} seconds, which is very short).

More traditionally, black holes are created by the collapse of massive stars. At the end of a star's life cycle, when the fuel inside the star has all run out, the mass of the star itself causes it to collapse, and all the matter inside of the star gets sucked in. Sometimes, this can create smaller, dimmer stars or create quasars (a type of ultra-bright, fast spinning, tiny star containing a black hole). Remember Heisenberg's uncertainty principle? Theoretically, this also can apply to the creation of black holes. If you will remember, according to the uncertainty principle, it is impossible to determine a quantum particle's momentum and position with high precision at the same time. If you attempt to measure one or the other more precisely, the different measurement becomes less precise. If you get enough accuracy— say, down to one Planck length or less—the additional value becomes so large that the particle is mathematically capable of turning into a black hole. It does not mean the particle is becoming a black hole, but more of a math problem illustrating the difficulty of quantum particles. According to physicists, it is called an absurdity or a nonsense result, which means that something is missing from the math. They are missing a formula, a variable, or something else. It is the mystery of quantum gravity—even today, we are still missing something in the math required to make gravity (and, by extension, black holes) make sense.

These issues are propagated because there is no way to measure gravity on a coordinate plane. Since gravity exists within four dimensions, it cannot be figured out using the two-dimensional math that scientists have traditionally used to figure out other quantum theories. Additionally, if you try to apply four-dimensional gravity to these two-dimensional theories, you receive anomalous answers—the math does not make sense. Essentially, gravity is space-time. As a result, you cannot anchor any math within space-time to figure it out.

At its core, this math does not work because Einstein's general theory of relativity is incompatible with quantum mechanics on a fundamental level. The general theory of relativity works at a level that we can see, but it unravels at the quantum level (i.e., below the Planck scale). It brings us full circle because that is why we created quantum mechanics in the first place. However, quantum mechanics has yet to account for the anomaly of gravity. For now, physicists continue to assume that we have not found the right theory to solve it, however.

According to Einstein's theory of general relativity, black holes are essentially anomalies in space-time that are holes with infinite mass. Black holes (and everything in the universe that experiences gravity experience gravitational time dilation to a lesser extent). The closer you are to the black hole center, the slower time moves to an observer outside the black hole. However, strangely enough, time will proceed at an average pace to those closest to the black hole. It means that, if a pair of twins were to stand by a black hole—one twin very close to the black hole, and the other very far away—when the twins both came back to Earth, the twin who was further from the black hole will have matured more than the one who was close to the black hole. Crazy, right? To a lesser extent, it also happens to the satellites orbiting the Earth, as well as the International Space Station (the ISS). Clocks that orbit Earth from space will slowly move ahead of the Earth's clocks.

Even stranger—continuing with the twin metaphor—if one of the twins were to fall inside the black hole, and the other twin was to watch, time would appear to completely stop on the twin inside the black hole as the twin crossed over the edge. The twin overlooking outside the black hole would see their other twin stop, freeze totally in time, then never move again, no matter how long the other twin waited for it. No one knows whether it would be possible to escape a black hole after this had happened (according to all known laws of physics, any ordinary person going into a black hole would be crushed by the powerful gravitational force long before they made it to the center, of course) if it were possible to survive the experience. One theory is the white hole theory, which says that for every black hole, there is a matching white hole somewhere else in the galaxy that spews out all of the matter consumed by the black hole. If the person who fell into the black hole survived, then according to this theory, the black hole would work as a teleportation device. However, although we have documented cases of proposed white spots in the universe (in

2006, the white hole GRB 060614 was found), there's no way to prove that these holes are in any way connected to black holes.

Some scientists have anticipated that the Big Bang created the universe as a white hole. The same paper proposes that white holes should be spontaneous, limited occurrences rather than long-lasting singularities like black holes.

Chapter 23. Real-World Applications

While physics does play a role in our lives, most of it involves things we don't think about. For example, physics helps to define how our world is put together on a molecular level. This understanding helped scientists to split atoms and use various waves to transmit information via data and sound.

At the same time, it's interesting to look at how brane physics, just one area, can be used to help us understand dimensions, even ones that might not be easily found or seen. Various aspects of brane physics have been used in cosmology. For example, brane gas cosmology attempts to clarify the three dimensions of space by topological and thermodynamic contemplations.

This idea puts forth three as the largest number of spatial dimensions because that is where strings can usually interconnect. Initially, there might be multiple windings of strings around dense dimensions, but space can only expand to macroscopic sizes once these windings are removed.

To remove these windings, oppositely wound strings must find each other and then be annihilated. But they can't find each other in only three dimensions, so this follows the idea that only three dimensions of space can grow larger with this initial configuration.

In just this one area, we have seen how quantum physics is working to understand how dimensions, space, and time work together in our universe. It is the greater understanding of how our universe works that ultimately brings quantum physics or mechanics to us.

But quantum physics also has a practical, everyday effect on our lives. For many, it used to reside on their wrists, but now it's part of our smartphones. Those incredibly precise timepieces are what we rely on to keep not only our schedule but also to keep our technology running—and they are based on quantum physics. So how does this translate to our clocks?

Let's start with the atomic clock. It is an incredibly precise piece of machinery that monitors and measures radiation frequencies and makes electrons jump from one energy level to another. This "fountain" clock, which uses cesium atoms and is housed in Colorado, will not gain or lose a second in 100 million years.

Now that is one accurate clock! These very sensitive clocks, in general, are part of GPS navigation, surveying, and even telecommunications. So even though quantum physics describes super tiny parts and pieces, it also plays a part in the technology of our everyday lives.

Researchers are looking at how to use quantum entanglement to make an even more precise clock. The reason this clock would be even more precise is that the atoms inside would not mark time by noting differences among their neighboring atoms but instead would work largely like a giant swinging pendulum. There is a hope that these clocks could be linked into a worldwide network and thus accurately measure time regardless of their location within the network.

Another area where quantum physics has helped to create a giant leap forward is in the area of supercomputers. These computers rely on quantum bits to speed up processing. While the field is still in development, progress is being made by scientists around the world as they attempt to harness quantum physics to make our computers run even faster.

As we have seen, quantum physics forms an aspect of the growth of technology and accuracy of measurement, but it is also growing in the contributions it makes to our society. The future of quantum studies is still relatively young, but as time goes on, so will be the incredible contributions it makes to the way we live and our understanding of how the world works.

In early 2015, scientists discovered an intermediate black hole in one of the spiral arms of the NGC 2276 galaxy, which they called

NGC2276-3c. This black hole appears to be about 50,000 times the sun's mass.

This intermediate-mass black hole (IMBH) may help to fill in some of the gaps in our knowledge about these exotic and amazing objects in space. It was first observed using x-rays and radio waves. NASA's Chandra X-ray Observatory combined optical data from the Hubble Space Telescope, a Digitized Sky Survey, and the European VLBI Network (EVN) to create a composite image of what it must look like and have mapped out NGC2276-3c.

By combining the data from the X-ray and radio waves, astronomers were able to make the educated guess that it is an intermediate-mass black hole. These black holes are larger than the typical stellar black holes (babies) and meaningfully smaller than the supermassive black holes (giants).

Researchers have estimated NGC2276-3c's mass based on a well-known and documented relationship between how bright the origin of the X-rays and radio waves are and the size of the overall mass of the black hole being studied. The brightness measurements are built on observations from the EVN and Chandra.

Astronomers have a great interest in IMBH because they are viewed as the beginning of what will become the giants or supermassive black holes. These IMBH also have a strong influence on their surrounding environment, including other stars and planets.

One of these ways is being demonstrated by NGC2276-3c, which appears to be constraining the creation of other stars within the neighboring area. The EVN documented data based on radio waves, which expose an inner jet that seems to have started six light-years away from NGC2276-3c. Additional observations have shown even larger-scale radio emissions expanding out over 2,000 light-years from their source, the black hole.

The jet's region extends roughly 1,000 light-years away from NGC2276-3c, and it appears to be without any young stars.

Scientists surmise that the jet has cleared a cavity within the gas disk surrounding the black hole and thus prevented new stars from forming within that area.

Yet, at the edge of the jet's radio emission, data has revealed a rather large star population. This enhanced star production could be taking place when the material the jet clears from the gas disk collides with other gas and dust found with the stars in NGC 2276. Another possibility is that new star production is ramped up when NGC 2276 merges with another dwarf galaxy.

Other studies of this galaxy, including observations made via Chandra, have taken time to examine the rich population of ultra-luminous X-ray sources (ULX) within this galaxy.

There are at least 16 X-ray sources that have been located. Of these, eight are considered ULXs. This group includes NGC2276-3c. One apparent ULX was found to be five separate ULXs, including NGC2276-3c. This study noted that anywhere from five to 15 solar masses worth of stars are being formed each year, just within the NGC 2276 galaxy.

So, what could be causing this high rate of star creation within this particular galaxy, despite the apparent suppression from NGC2276-3c's jet? Scientists believe this high rate of star creation has been triggered by an impact with a dwarf galaxy, which would seem to support the original theory of mergers that create supermassive black holes (giants).

Scientists will continue to study these black holes in hopes of coming to a better understanding of how black holes are created in terms of their size, but also to understand their life cycle. This means building a foundation of knowledge on how they mature and how their destruction occurs.

As scientists continue to study objects such as NGC2276-3c, they will look for evidence of a point when a black hole grows so big that it can destroy itself or even splinter back into smaller black holes. The amount of energy captured within a black hole could mean a large energy release upon their destruction or death.

Studies will continue as this unique part of space beckons to our collective imagination.

Conclusion

Thank you for making it through to the end of this book.

Quantum physics is an endlessly fascinating subject and one that we think everyone should learn about in some capacity. After all, as we discover more and more about quantum physics, more of the future is open to us-things like teleportation, supercomputing, and ultra-fast space travel all feel like they're just around the corner if we can finally unravel how quantum mechanics works. We're not all scientists, of course, and we can't all conduct experiments to figure out the next big discovery, but we can all do our part just by learning a little bit about how the universe around us works.

There are so many things going on in the universe around us that we can't explain. However, there's beauty in this un-knowing, and there's an indelible fascination in the discovery of the world around us. How boring would it be to know already exactly how everything around us worked? If we knew that, we would be able to predict everything, from whether there was alien life in the far reaches of the universe to when our planet and solar system would die. While some might argue that it might be better to know, the vastness of the universe allows for infinite new things to discover, crazy new laws and theorems to figure out, and new phenomena just over the event horizon.

The next step is to incorporate the knowledge you've gained here into your everyday life. Whether you use your new knowledge simply to brag, to help others, to do your research, to deepen your understanding of our world, or to become the next Albert Einstein, we're sure it will enrich your life in new and exciting ways. Please never stop reading, and more importantly, never stop learning.